T0231414

Cardiothoracic Surgery

Third Edition

Fritz J. Baumgartner, M.D.
Clinical Assitant Professor of Surgery
UCLA School of Medicine
Division of Cardiothoracic Surgery
Harbor-UCLA Medical Center
Vascular Surgery Associates
Long Beach, California
U.S.A.

CRC Press
Taylor & Francis Group
Boca Raton London New York

CRC Press is an imprint of the
Taylor & Francis Group, an **informa** business

VADEMECUM
Cardiothoracic Surgery Third Edition

First published 2004 by Landes Bioscience

Published 2019 by CRC Press
Taylor & Francis Group
6000 Broken Sound Parkway NW, Suite 300
Boca Raton, FL 33487-2742

ISBN 13: 978-1-57059-683-4 (pbk)

Visit the Taylor & Francis Web site at
http://www.taylorandfrancis.com

and the CRC Press Web site at
http://www.crcpress.com

Library of Congress Cataloging-in-Publication Data

Baumgartner, Fritz J., 1957-
  Cardiothoracic surgery / Fritz J. Baumgartner.-- 3rd ed.
     p. ; cm. -- (Vademecum)
  Includes index.
  ISBN 1-57059-683-2
  1. Heart--Surgery. 2. Chest--Surgery. I. Title. II. Series.
  [DNLM: 1. Cardiac Surgical Procedures--methods. 2. Thoracic Surgical Procedures--methods. WG 169 B348c 2004]
  RD598.C37 2004
  617.4'12--dc22   2003028082

Dedication

Dedicated to my mother and father,
wife and baby, with love and thanks.

Contents

Editor

Fritz J. Baumgartner, M.D.
Clinical Assistant Professor of Surgery
UCLA School of Medicine
Division of Cardiothoracic Surgery
Harbor-UCLA Medical Center
Vascular Surgery Associates
Long Beach, California
Chapters 1-3 and 5-25

Contributors

Matthew Budoff, M.D.
Assistant Professor of Medicine
UCLA School of Medicine
Program Director
Division of Cardiology
Harbor-UCLA Medical Center
Chapters 2, 5

John R. Charney, M.D.
Chief, Division of Cardiothoracic Anesthesia, Harbor-UCLA
Professor of Anesthesiology, UCLA School of Medicine
Associate Clinical Assistant Professor of Pediatrics,
UCLA School of Medicine
Chapter 4

Bassam O. Omari, M.D.
Chief, Division of Cardiothoracic Surgery, Harbor-UCLA
Assistant Professor of Surgery, UCLA School of Medicine
Consultant Staff, St. John's Hospital
and Health Care Center, Santa Monica
Chapters 2, 8, 10

John R. Robertson, M.D.
Chief, Thoracic and Cardiovascular Surgery
St. John's Hospital and Health Care Center, Santa Monica
Clinical Assistant Professor of Surgery, UCLA School of Medicine
Chapter 10

Preface

"The hospital, the operating room and the wards should be laboratories, laboratories of the highest order."

—*William Halsted*

Extensive changes have been incorporated into *Cardiothoracic Surgery, Third Edition,* but as with the prior two editions, the Third edition continues as a manual intended for residents at any stage of their experience, including cardiothoracic fellowship. It is meant as a guideline to understand the pathophysiologic rationale of what we do in the operating room and intensive care unit. Emphasis is placed on physiology, anatomy, pathology, medical management, and surgical technique. It is not meant to be fully and completely comprehended on the first reading, but will hopefully succeed in making the experience of cardiothoracic surgery more understandable and enjoyable. Above all, the welfare of the patient must be the ultimate goal of any such endeavor.

As physicians, we took a vow to preserve life, often at great personal sacrifice. That vow is integral to the Hippocratic Oath, which, in its unadulterated version, states: "I will not give to a woman an instrument to produce abortion. With purity and holiness I will pass my life and practice my art." Human life is sacred, beginning in the womb and ending in the grave, a fact society and medicine sometimes ignore to their grave detriment.

In a world where self-gratification and freedom of choice are often substitutes for discipline and compassion, we take solace in the observations of the great neurosurgeon Harvey Cushing:

"...Only when the gift requires self-denial and only if the giver... speaketh the truth in his heart, will he, like St. Francis, come to be... forever blessed."

Between two people of equal technical skill, the one who cares will do the better job.

—*Fritz J. Baumgartner*

Acknowledgments

The principles of patient care have been passed down over generations of cardiothoracic surgeons at Harbor-UCLA. Drs. Jeffrey Milliken, John R. Robertson, and Ronald Nelson are gratefully acknowledged for entrusting their skills to me and future generations of surgeons.

Anatomy

Fritz J. Baumgartner

CHAMBERS

The anatomic structures of the right atrium of importance include the fossa ovalis which is the obliterated ostium secundum. This is surrounded by a limbus or raised area of tissue which encircles the fossa ovalis at all but its inferior margin. There is a wide-based, blunt atrial appendage. The Eustachian valve exists as a flap at the orifice of the inferior vena cava and the thebesian valve similarly exists at the orifice of the coronary sinus. There is a crista terminalis which surrounds the base of the right atrial appendage and thus separates the trabeculated from the nontrabeculated portion of the atrium.

The left atrium receives pulmonary venous drainage usually from right and left superior and inferior pulmonary veins which drain into the posterior portion of the left atrium. Unlike the blunt right atrial appendage, the left atrial appendage is long, narrow, and fern-shaped, and there is no crista terminalis in the left atrium. Unlike the right atrium, the left atrium is not trabeculated. The left atrial appendage is the only trabeculated structure in the left atrium.

The right ventricle is usually located anteriorly and to the right side of the left ventricle. The inflow of the right ventricle is via the tricuspid valve. Flow enters the right ventricle into a large sinus portion and a smaller infundibulum or outlet portion just proximal to the pulmonary valve. The sinus and infundibular portions of the right ventricle are coarsely trabeculated. This is the major distinction between the right ventricle and the left ventricle, i.e., the right ventricle is much coarser than the smoother left ventricle, and this is an important differentiation factor on ventriculography.

A septum lies between the inflow and outflow portions of the right ventricle and thus lies adjacent to the pulmonary valve. This septum is called the infundibular septum or the conal septum or the crista supraventricularis. These are synonymous terms and the function of this septum is to separate the pulmonary valve from the confluence of the aortic valve, mitral valve and tricuspid valve. The fusion of the latter three structures forms the fibrous skeleton of the heart.

The tricuspid valve has three leaflets, and its anatomy will be described later. The papillary muscles which support these three leaflets differ in basic anatomy from the papillary muscle arrangement of the mitral valve in the left ventricle. There is a single large anterior papillary muscle attached to the anterior free wall in the case of the tricuspid valve. There are multiple smaller posterior papillary muscles attached partly to the free wall of the ventricle and partly to the interventricular septum.

The left ventricle consists of a large sinus portion or inflow portion which includes the mitral valve and the apex and a much smaller outflow portion beneath the aortic valve. The inflow and outflow portions of the left ventricle are separated by the anterior leaflet of the mitral valve. Although the wall of the left ventricle is trabeculated, the trabeculae are fine compared to the coarse trabeculations in the right ventricle. The outflow portion of the left ventricle lies anteriorly and to the right of the anterior leaflet of the mitral valve. It should be noted that this outflow portion of the left ventricle abuts the inflow portion of the right ventricle, the two being separated by the interventricular septum. The anterior leaflet of the mitral valve attaches to the interventricular septum as well as a portion of the aortic valve. This will be described further later on.

On the right ventricular side, only the septal leaflet of the tricuspid valve attaches to the septum, whereas on the left ventricular side both the anterior leaflet of the mitral valve as well as a portion of the aortic valve attach to the interventricular septum. The left half of the anterior leaflet of the mitral valve is in fibrous continuity with the aortic valve in an area called the aortic mitral annulus. This is shown in Figure 1.1 and will be described later. This is the most important figure in understanding valvular cardiac anatomy and what the implications are for suture placement during valve replacement in terms of juxtaposing structures. The papillary muscle anatomy in the left ventricle is generally uniform and consists of two large papillary muscles attached to the free wall. These are an anterolateral papillary muscle and a posteromedial papillary muscle. Both of these attach to the free wall of the left ventricle unlike the right ventricle where some papillary muscles attach to the interventricular septum. The anterolateral and posteromedial papillary muscles attach to both the anterior and posterior leaflets of the mitral valve. The anterolateral papillary muscle is less prone to ischemia than the posteromedial papillary muscle because the anterolateral papillary muscle is supplied by septal branches of the left anterior descending artery as well as the circumflex artery. The posteromedial papillary muscle is supplied by only the right coronary artery.

With regard to the interventricular septum, the right and left ventricular septal surfaces are asymmetric due to the presence of the infundibulum in the right ventricle only. Also, the higher pressure in the left ventricle makes the left ventricular septal surface concave with respect to the convex right ventricular surface.

The axis of the right and left ventricular outflow tracts are also different with the RV outflow tract being directly superiorly oriented while the LV outflow tract is angled towards the right. The atrial-ventricular septum separates the atrium from the ventricle and consists of a membranous AV septum and a muscular AV septum. The membranous AV septum is associated with the fibrous skeleton of the heart. Directly between the membranous and muscular AV septum lies the AV node and conduction tissue.

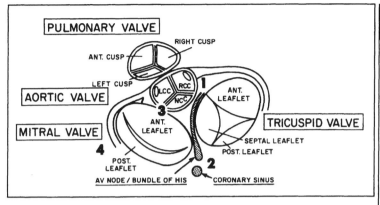

Fig. 1.1a. Anatomic valvular relationships and hazardous surgical areas. 1) Bundle of His: vulnerable just under junction of right and noncoronary cusps of the aortic valve (RCC and NCC). It is also found coursing adjacent to the septal leaflet of the tricuspid valve. It is particularly vulnerable at the junction of the septal and anterior leaflets of the tricuspid valve. 2) Atrioventricular node: vulnerable in the region between the coronary sinus and tricuspid valve annulus. Also found at the 2 o'clock position of anterior leaflet of mitral valve. 3) Junction of the aortic left and noncoronary cusp (LCC and NCC): These two cusps may sustain injury when suturing at the 10 o'clock position on the mitral annulus. 4) Circumflex artery, coronary sinus, and left atrioventricular groove: may be injured by deep sutures at the 7 o'clock position of the mitral annulus.

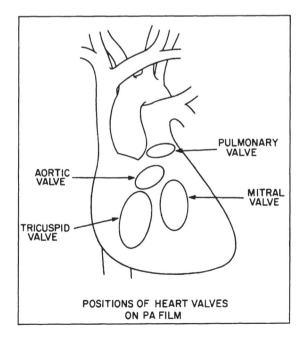

Fig. 1.1b. Relative positions of heart valves on a PA chest film.

THE CONDUCTION SYSTEM

The sinus node is located on the anterolateral aspect of the junction between the superior vena cava and right atrium. It can sometimes be seen as a white nodule in this area. It is superficial, lying just beneath the epicardial surface. The AV node lies on the right atrial side of the central fibrous body (right trigone) between the muscular and membranous portions of the AV septum. It is just anterosuperior to the ostium of the coronary sinus in the triangle of Koch (Fig. 1.2). The triangle of Koch is formed by the tricuspid annulus, the coronary sinus, and the tendon of Todaro (a continuation of the eustachian valve from the inferior vena cava). The bundle of His passes through the right part of the right trigone and the central fibrous body, and this area is just inferior to the commissure between the septal and anterior leaflets of the tricuspid valve. It is this area which is in most danger of injuring conduction tissue when stitches are placed in the tricuspid annulus. The bundle of His then courses along the border of the membranous septum giving out fibers forming the left bundle branch. This region of the junction of the septal and anterior leaflet of the tricuspid valve is exactly the location between the right coronary cusp of the aortic valve and noncoronary cusp of the aortic valve in the anterior commissure and it is precisely this region which is in most danger of injuring the bundle of His when performing an aortic valve replacement (Fig. 1.1).

The left bundle branch fans over the left ventricular septum surface and is subdivided into anterior and posterior subdivisions, the anterior going to the anterolateral papillary muscle and the posterior going towards the posteromedial papillary muscle.

The right bundle branch originates from the bundle of His and fans out to supply the wall of the right ventricle.

THE CARDIAC VALVES

The interrelationship between the aortic, pulmonary, mitral and tricuspid valves is quite uniform. The pulmonary valve is somewhat anterosuperior and to the left of the aortic valve. The annulus of the aortic, mitral, and tricuspid valves merge with each other and the membranous septum to form the fibrous skeleton of the heart. The anterior leaflet of the mitral valve is in fibrous continuity with portions of the left and noncoronary aortic cusps. This skeleton has left and right fibrous trigones. The right fibrous trigone is the junction between the mitral, tricuspid, and aortic annuli and the membranous septum, and is pierced by the bundle of His; as noted above this is the most dangerous place for injuring conduction tissue when performing a mitral, aortic, or tricuspid valve replacement. The left fibrous trigone is situated more to the left and lies between the left aortic cusp and mitral annulus.

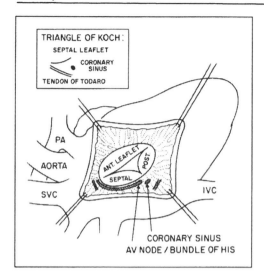

Fig. 1.2. Interior of the right atrium as seen from the surgeon's view.

MITRAL VALVE

The mitral valve is bicuspid with an anterior (septal) leaflet and a posterior (mural) leaflet. Figure 1.1a shows that the anterior mitral leaflet has a much larger area than the posterior leaflet, but the circumference of the posterior leaflet is much larger than the anterior leaflet, hence the smile configuration of mitral valve. The septal or anterior leaflet is in fibrous continuity with the aortic valve through the aortic mitral annulus as described above. The region of continuity occupies about one quarter of the mitral annulus and corresponds to the region between half of the left coronary cusp and half of the noncoronary cusp of the aortic valve (Fig. 1.3).

The limits of this attachment are demarcated by the right and left fibrous trigones. These points do not correspond to the commissures of the mitral valve, although the commissures are close. The AV node and bundle are at risk of surgical damage because of the proximity to the right trigone which is adjacent to the right and noncoronary cusp of the aortic valve and to the septal and anterior leaflet of the tricuspid valve.

The posterior leaflet, although smaller, occupies more of the circumference of the mitral valve annulus. The chordae tendineae to the mitral valve originate from the anterolateral and posteromedial papillary muscles. Each leaflet receives chordae from both muscles with the majority inserting on the free leaflet edge. There are three orders of chordae: the first order insert on the free margin of the leaflet; the second order chordae insert several millimeters back from the free edge; and the third order chordae insert at the base of the leaflet. Third order chordae exist only on the posterior leaflet.

TRICUSPID VALVE

The annulus of the tricuspid valve is relatively indistinct, compared with the mitral valve, especially in the region of the septal leaflet. The leaflets and chordae are thinner than that of the mitral valve. The anterior leaflet is the largest of the leaflets; the posterior leaflet is usually smallest; the septal leaflet is larger than the posterior leaflet. Of major importance is the proximity of the septal and anterior leaflet commissure to the membranous septum immediately adjacent to the bundle of His which penetrates the right trigone in this region.

AORTIC VALVE

This is a tricuspid valve consisting of a right, left and noncoronary cusps. The aortic valve is in fibrous continuity with the anterior leaflet of the mitral valve and the membranous septum. The walls of the coronary sinuses are thinner than the wall of the remainder of the aorta itself.

PULMONARY VALVE

The structure of the pulmonary valve is similar to the aortic valve. The pulmonary valve has three cusps which are lighter than that of the aortic cusps.

Pulmonary valve cusps are described by several terminologies, but usually by their relationship to the aortic valve, i.e., right, left and anterior pulmonary cusps.

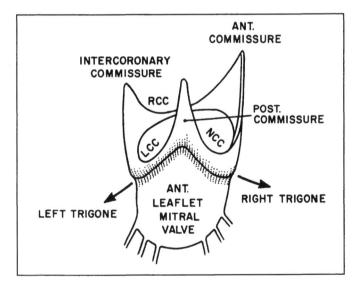

Fig. 1.3. Anatomic relationship of the anterior leaflet of the mitral valve and the aortic valve.

CORONARY ARTERIES

These consist of a left main coronary artery which bifurcates to the circumflex coronary artery and left anterior descending coronary artery (Fig. 1.4). The orifice of the left main coronary artery lies in the left coronary sinus. The orifice of the right coronary artery arises from the right coronary sinus. The circumflex coronary artery lies in the groove between the left atrium and left ventricle adjacent to the coronary sinus. The circumflex coronary artery gives off obtuse marginal branches because they lie on the obtuse surface of the heart. The branches of the left anterior descending artery are called diagonal branches. The right coronary artery has a first branch called an acute marginal artery which supplies the free ventricular surface and may also give off a conal branch which supplies the infundibulum of the right ventricle. The right coronary artery continues to bifurcate into a posterior descending artery as well as a continuing posterior ventricular branch, also called a posterolateral branch or LV extension branch. The dominance of the coronary arteries is determined by which side, i.e., right or left, supplies the posterior descending artery. In 90% of people, the posterior descending artery is a continuation of the right coronary artery; in 10% it is a continuation of the circumflex coronary artery or the left anterior descending coronary artery. The Collaterals going from the right coronary artery to the left anterior descending artery comprise the loop of Vieussens. Septal arteries arise perpendicularly from the left anterior descending artery to supply the interventricular septum. This is often helpful in angiographic identification of the LAD and in differentiating the septal branches from diagonal branches. The right coronary artery courses down the right atrioventricular groove. The right coronary artery crosses the crux

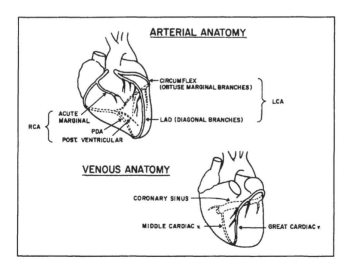

Fig. 1.4. Arterial and venous anatomy of the heart.

of the heart where it gives off an AV node artery at the apex of the bend. It then terminates into the posterior descending artery and posterolateral segment artery (same as posterior ventricular branch or LV extension branch).

The sinus node artery usually originates from the right coronary artery. The sinus node artery arises from the right coronary artery in 55% of hearts and from the left circumflex or left main coronary artery in the rest. As noted above, the AV node artery usually arises from the bend of the right coronary artery as it crosses the crux of the heart. The AV node is usually supplied by the dominant coronary artery.

An important accessory blood supply to the AV node is Kugel's artery which originates from the right coronary artery or circumflex artery, coursing through the interatrial septum. The left ventricular anterolateral papillary muscle is supplied by the left anterior descending artery and obtuse marginal branches; however, the posteromedial papillary muscle is supplied by the terminal right coronary artery or terminal circumflex coronary artery, i.e., whichever is dominant in the particular person. The posteromedial papillary muscle is supplied by only one coronary artery. The posteromedial muscle is thus more often susceptible to ischemia in patients with coronary artery disease than the anterolateral papillary muscle. This is of clinical significance because ischemia of the posteromedial papillary muscle may lead to ischemic mitral regurgitation. In more severe cases the muscle may infarct, leading to rupture of the papillary muscle, severe mitral regurgitation and pulmonary edema.

SUGGESTED READING

1.   Wilcox BW, Anderson RH. Surgical anatomy of the heart. In: Baue AE, Geha AS, Hammond GL et al, eds. Glenn's Thoracic and Cardiovascular Surgery. 5th ed. East Norwalk: Appleton and Lange, 1991:1333-1344.
2.   Wilcox BR, Anderson RH. Surgical Anatomy of the Heart. New York: Raven Press, 1985.
3.   McAlpine WA. Heart and Coronary Arteries. New York: Springer-Verlag, 1975.

Cardiac Diagnosis

Matthew Budoff, Fritz J. Baumgartner and Bassam O. Omari

ECHOCARDIOGRAPHY

Echocardiography has become an exceedingly useful study in the noninvasive evaluation of cardiac disease. It is biologically safe and can therefore be repeated often without concern for exposure. Real-time two-dimensional imaging provides data on ventricular size and function, including both global and regional wall abnormalities. It also provides information on valve morphology and function such as leaflet prolapse or leaflet motion abnormalities, or the presence of masses or vegetations. Adding Doppler velocities provides specific information in evaluating the severity of valvular disease including pressure gradient and valve area. Most valvular heart disease requiring surgical intervention can be adequately diagnosed and assessed on the basis of history, physical exam and echocardiogram.

Although transesophageal echocardiography is invasive, it can be used as a complement to transthoracic echocardiography in the preoperative evaluation. It provides superior images of the cardiac structures due to the proximity of the probe to the heart. It is also very useful in the intraoperative and postoperative evaluation of these patients.

In mitral valve disease, the echocardiogram can demonstrate any rheumatic disease or changes such as thickening of the leaflets, fusing of the commissures or shortening of the chordae. It can also demonstrate chordae rupture, leaflet prolapse, or leaflet perforation. Occasionally, it may show lack of complete leaflet coaptation. Echocardiography can be used to assess the size of the left atrium and the presence or absence of clot in the left atrial appendage. In the case of mitral regurgitation, it can be used to grade the degree of regurgitation. Mild regurgitation (1+) is when the regurgitant flow is located only near the valve, while severe regurgitation (4+) is when the regurgitant flow extends to the pulmonary veins.

In aortic valve disease, the echocardiogram can demonstrate whether the valve is bicuspid and shows any annular or leaflet calcifications, or any leaflet motion abnormalities. In the case of aortic stenosis, the echocardiogram can determine the pressure gradient across the valve and the valve area; and in the case of aortic insufficiency, it can assess the severity of the regurgitation.

The ability to measure the aortic root helps to gauge the appropriateness of different options for surgery, including the size of the prosthetic valve. Valve leaflet motion abnormalities and calcification can also be visualized with echocardiography. Similar to mitral regurgitation, the presence and severity of aortic valve regurgitation can be determined and graded on a scale of 1-4. Similarly, it is possible to measure the presence and severity of pulmonic and tricuspid valve regurgitation and stenosis.

In atrial fibrillation the size of the atrium, presence of clot and associated ventricular function can be assessed. Transesophageal echocardiography is now standard prior to cardioversion to rule out the presence of atrial clot, including clot in the left atrial appendage. Estimates of pulmonary pressure, based upon velocity and underlying pressure, can allow for the noninvasive monitoring of many cardiac diseases, including pulmonary hyperetension. This also allows for serial evaluation in order to time the need for valve surgery.

Echocardiography is often the first test employed for cases of congenital heart disease. Echocardiography has the ability to visualize and semi-quantitate cardiac shunts (including atrial septal and ventricular septal defects, anomalous pulmonary venous return and patent ductus arteriosus), as well as other complex heart diseases. Echocardiography allows for evaluation after myocardial infarction to determine infarct size, residual ejection fraction as well as post-infarct complications such as ventricular rupture, pericardial effusion, aortic dissection, new mitral regurgitation and ventricular septal defects. The significant limitation of echocardiography is incomplete visualization of the endocardium; because of this the ability to quantitate ejection fraction and wall motion is somewhat limited. The use of new contrast agents (given intravenously during ultrasound exam) has dramatically improved the performance of echocardiography to clearly visualize the endocardium (and perhaps to roughly measure perfusion), although it still remains a somewhat subjective measure, limited by high inter-reader variability.

Exercise echocardiography improves upon the sensitivity and specificity of stress echocardiography by allowing for rest and stress imaging of the myocardium. The presence of a stress-induced wall motion abnormality is consistent with a highly stenotic coronary lesion. Stress testing allows for assessment of exercise capacity as well as stress-induced ischemia (chest pain or electrocardiographic changes consistent with ischemia) with a sensitivity of approximately 70%. Stress echocardiography, by visualizing the wall during rest and increasing levels of stress, increases the sensitivity and specificity of treadmill tesing to approximately 85-90%. Exercise often is done on a bicycle or treadmill. For patients who cannot exercise (those with poor exercise tolerance or other limitations such as arthritis), dobutamine is most often administered to increase myocardial demand. The relative change in wall motion from rest to stress, whether by exercise or drug, allows for assessment of myocardium that becomes underperfused. Resting abnormalities of a certain wall would be consistent with infarct, while new abnormalities that develop would signify ischemia. While more expensive ($800-$1,000) than stress testing, which usually averages about $300, it is considerably less invasive and expensive than radionuclide imaging and can be performed in approximately 30 minutes.

RADIONUCLIDE IMAGING

Nuclear cardiac imaging procedures are noninvasive techniques that provide valuable information for cardiac disease in two main broad categories. The first category involves the assessment of ventricular function by radioactive tracers that remain in the intravascular space during the study. The second category of imaging visualizes myocardial intracellular uptake of radioactive tracers reflecting myocardial perfusion, metabolism and viability.

The multigated equilibrium blood pool imaging study is used to assess the cardiac performance. This study is done using the patient's own red blood cells tagged with Technitium-99m (^{99m}Tc). This method provides a relatively stable blood pool for up to 6 to 8 hours with a biologic half-life of approximately 4 hours. This technique is used to determine the left ventricular ejection fraction as well as any wall motion abnormalities.

Thallium-201 is a potassium analog taken up by viable myocardium. Studies using Technetium-99 and Thallium-201 include wall motion evaluation, myocardial perfusion and viability assessment. The initial distribution of these agents at rest demonstrates the areas of adequate blood flow and viable tissue. Both underperfused and infarcted tissue will appear dark (cold spot) on the resultant images. The ability to reimage several hours later will sometimes demonstrate viable myocardium, where initially blood flow was inadequate, but over time there is enhancement of these regions. Studies have used both dobutamine and nitroglycerine to enhance the filling, further demonstrating the presence of viable myocardium. Presumably these regions are chronically ischemic or hibernating areas of myocardium that may benefit from revascularization.

Exercise radionuclide imaging is the most common application of this technique, especially in patients with suspected coronary artery disease (CAD). These patients have normal resting images (good perfusion to all areas of the myocardium) but decreased uptake (dark or cold spots) after exercise. This demonstrates regions of inadequate blood flow during exertion (perfusion defects) diagnostic of obstructive coronary artery disease. The published sensitivity and specificity of these techniques, as compared to cardiac catheterization to diagnose obstructive CAD, is approximately 85%. A reversible defect (one not present on the resting images but present on exertion) is synonymous with obstruction, while a fixed defect (one present on both exercise and rest) represents infarcted tissue.

Patients who cannot achieve adequate exercise levels are often given agents that will cause differential blood flow during rest and stress. Intravenous persantine is most often used to mimic regional ischemia and allow detection of obstructive coronary artery disease. Persantine is a potent vasodilator that causes selective coronary hyperemia. A differential rate of radionuclide uptake from rest to stress makes the diagnosis. The basis of this test is that a stenotic lesion cannot vasodilate so there is more blood flow to normal regions of the heart, and the myocardium distal to the stenotic lesion appears less perfused (dark). An alternative to this is to use dobutamine or arbutamine stress to increase cardiac requirements (see Echocardiography, above), thus allowing relative hypoperfusion in areas distal to stenoses.

The limitations of radionuclide testing are the need for intravenous injection, high cost (approximately $2,000 in most laboratories), radiation exposure, as well as a slow imaging protocol. The images take up to 4 hours to obtain, requiring the patient remain immobile. Furthermore, often the rest and stress imaging cannot be done on the same day so the patient must spend two days in testing.

CARDIAC CATHETERIZATION

Diagnostic coronary angiography is the standard of reference for determining the severity of stenosis resulting from atherosclerosis in the coronary arteries. Visualization of the lumen of the coronary artery to assess for the need for vascularization is performed by injection of iodinated contrast directly into the coronary arteries. Since the contrast resolution is limited with fluoroscopy (the imaging method used during angiography), direct enhancement of the blood pool in the artery of interest is required. Conventional coronary angiography (CCA) further

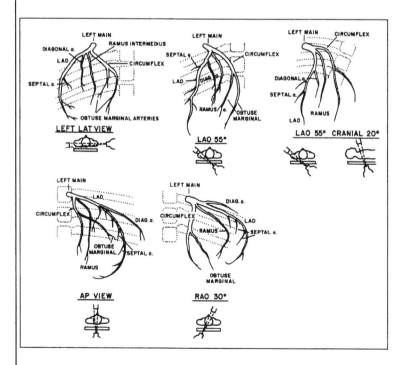

Fig. 2.1a. Cineangiographic views of left coronary artery.

requires direct arterial puncture and selective cannulation of the ostia of the left and right coronary arteries and, if present, each saphenous vein graft or internal mammary graft to obtain optimal selective contrast enhancement. Most commonly the catheter is threaded through the femoral or radial arteries, then retrograde up the aorta, and the coronary arteries are cannulated. Dye is injected and images are taken in several standard positions (Fig. 2.1a-b). The catheter also is passed retrograde past the aortic valve and into the left ventricular cavity to measure pressures as well as perform contrast ventriculography to assess wall motion, ejection fraction and the presence and severity of mitral regurgitation. The left ventricular silhouette on the RAO angiogram has five segments: anterobasal, anterolateral, apical, diaphragmatic and posterobasal (Fig. 2.2). Both akinesis and hypokinesis can be identified, as well as aneurysmal segments.

Coronary angiography identifies and assesses the severity of coronary artery atherosclerotic lesions. Generally the angiographic assessment of arterial stenosis can be divided into three major categories. The first group includes vessels with less than 50% luminal diameter narrowing, the second group includes vessels with 50-70% decrease in luminal diameter and the third group includes vessels with greater than 70% decrease. Reduction of the coronary artery diameter by 50% or

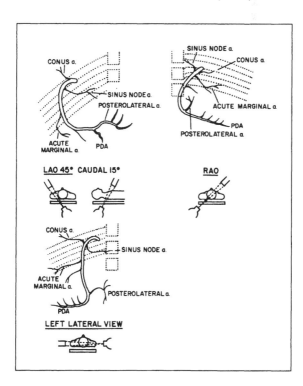

Fig. 2.1b. Cineangiographic views of right coronary artery.

Fig. 2.2. Right anterior oblique left ventriculogram.

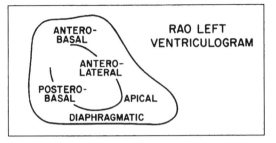

more is generally considered significant since it represents a 75% reduction in cross-sectional area. A 75% decrease in luminal diameter results in 95% reduction in cross-sectional area.

By selective catheterization of the right coronary artery and the left coronary artery and by using multiple views and projections, all the coronary arteries and their branches should be visualized. Lesions in these vessels can be identified and their severity can be assessed.

Angiography is done also to evaluate valvular heart disease. Exact ejection fraction is measured on ventriculography at cardiac catheterization by measuring the end diastolic volume minus the end systolic volume divided by the end diastolic volume. Normal ejection fraction is greater than 55%. The valve area can be measured according to the Gorlin formula by the mean gradients measured at cardiac catheterization. The valve area equals flow divided by 44 times the square root of the mean gradient across the valve. If both aortic insufficiency and mitral regurgitation are present, neither the aortic nor mitral valve area can be measured accurately at catheterization because the flow across a single valve cannot be assessed. The normal mitral valve orifice area in adults is 4 to 5 cm², and the normal aortic valve area is 2.6 to 3.5 cm².

The following valve gradients are considered severe stenosis:
1. aortic 50 mmHg
2. mitral 15 mmHg
3. tricuspid 5 mmHg
4. pulmonary 15 mmHg

When grading regurgitation of the mitral or tricuspid valve, 1+ indicates wisps of regurgitation going back into the left atrium or right atrium; 2+ regurgitation goes back more than half the diameter of the atrium; 3+ regurgitation hits the back wall of the left or right atrium; and 4+ regurgitation actually extends retrograde into the pulmonary vein or cavae. Generally, 1+ is mild regurgitation, 2-3+ is moderate regurgitation, and 4+ is severe regurgitation.

At cardiac catheterization, pressure time measurements can be performed with the catheter in the left ventricle and in the aorta to measure different conditions of aortic stenosis or aortic insufficiency.

Some therapeutic valvular interventions can be performed at the time of cardiac catheterization. Valvuloplasty can be performed in some instances, for example pulmonary valvuloplasty has proven to be effective for pulmonary stenosis. For mitral stenosis, valvuloplasty is considered in certain situations. Patients who have severe left ventricular dysfunction may undergo valvuloplasty as a bridge to later mitral valve replacement. Also patients of an extremely advanced age may be considered candidates for valvuloplasty, if their condition requires it. The results and chance of a stroke after valvuloplasty are similar to a closed commissurotomy (i.e., a commissurotomy which is done off cardiopulmonary bypass simply by opening the left atrium and placing the gloved finger through the mitral valve). One needs to keep the patient on coumadin for 6 weeks prior to valvuloplasty to prevent thrombus from forming in the atrium in the region of the valve. Valvuloplasty in aortic stenosis is less effective because of the chance of severe aortic insufficiency after the valvuloplasty.

In 1993, 1.8 million cardiac catheterization procedures were performed. Given the trend of increased utilization, it is possible by 2010 that annual use will exceed 3.0 million cardiac catherizations.

Coronary angiography is the only method currently available for defining the details of the entire coronary endoluminal vascular anatomy, and it provides the reference standard against which other tests are compared. Although coronary lesions that reduce luminal diameter <50% are considered hemodynamically insignificant, they are not clinically benign. These lesions can progress, either acutely or chronically, and patients with nonsignificant obstructions have significantly more cardiovascular events during follow-up than those with truly normal coronary angiograms.

CT CARDIAC IMAGING

CT, with most of the available validation and clinical data being derived from studies of electron beam tomography (EBT), is a robust methodology to evaluate coronary anatomy, vessel patency, atherosclerotic burden and ejection fraction both pre- and post-infarction.

EBT is a fourth generation CT imaging process, able to obtain thin slices of the heart and coronary arteries to evaluate anatomy. Rapid image acquisition due to the absence of a moving x-ray source allows approximately five times greater imaging speed than conventional multislice computed tomography (MSCT), limiting the respiratory and cardiac motion artifacts inherent in cardiac imaging. Usually 30-40 axial images are obtained to include the full-length of the myocardium.

The cine (or movie) scanning mode is designed to assess cardiac function. The scanning frequency of 17 scans/sec is sufficient to study both systolic and diastolic function. The spatial resolution adequately defines the endocardium of both the right and left ventricles so that precise measurements of cardiac volumes, mass and ejection fractions are feasible. Quantitative measurement of wall motion and wall thickening can be performed; it is particularly useful for evaluating CAD patients. Bicycle exercise can be coupled with EBT scanning to detect exercise-induced ischemia. Data indicates that exercise CT may be at least as sensitive and more specific than Technetium-99m sestamibi stress testing.

CORONARY ARTERY CALCIFICATION AND ATHEROSCLEROSIS

Calcific deposits in coronary arteries are pathognomonic of atherosclerosis. Detection of coronary calcium by EBT has been demonstrated to be highly sensitive for the presence of significant CAD, with a very high sensitivity and negative predictive value (>99%). Thus a score of 0 (no coronary calcium) can virtually exclude those patients with obstructive CAD. The most powerful and important data for this modality relates to its ability to predict future coronary events in both symptomatic and asymptomatic persons. Patients with increased levels of coronary calcium are at up to 10 times increased risk of future cardiac events.

NONINVASIVE ANGIOGRAPHY

Electron beam angiography (EBA) is a technology with the potential for obtaining essentially noninvasive coronary arteriograms. This procedure requires intravenous contrast to opacify the lumen and is completed within 20-30 minutes. Recent studies have reported this modality could be used to identify significant coronary lumen narrowing (>50% stenosis) with a sensitiivty of 92% and specificity of 94% as compared to invasive coronary angiography. Selective use of EBA might prove cost effective and provide a safer, less invasive method to assess luminal stenosis.

EBA in patients post-CABG or post-stent demonstrated sensitivities of 92-100% and specificities of 91-100% for establishing patency as compared to conventional coronary angiography (Fig. 2.3).

SPIRAL CT

EBT is not widely available, and the high cost of equipment prohibits widespread use. MSCT has been shown to have most of the capabilities of EBT, with promise of wider availability. Unfortunately, the speed of acquisition (temporal resolution) distinguished EBT from the results of the slower images from spiral CT, and hence spiral CT has not yet proved to be a feasible alternative to electron-beam CT for coronary artery calcification quantification. Multislice CT has the new capability of obtaining multiple images simultaneously, continuously imaging, obtaining 4, 8 or even 16 slices at once. Each slice is still obtained with the limited gantry rotation speed, but simultaneous imaging allows for thinner slices and shorter scanning protocols (decreased need for long breathholding by the patient).

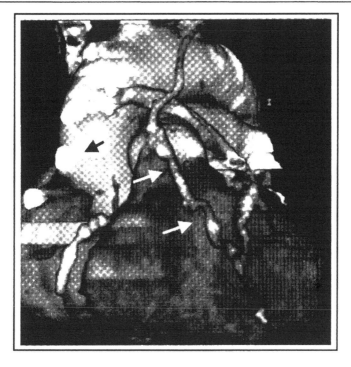

Fig. 2.3. A noninvasive angiogram demonstrating a widely patent left internal mammary artery to the left anterior descending (white arrows) and a closed saphenous graft to the distal right coronary artery (black arrow).

MAGNETIC RESONANCE IMAGING (MRI)

MRI provides excellent soft tissue contrast, has inherent 3-D capabilities and allows acquisition in any anatomic plane. Furthermore, MRI does not expose the patient to radiation nor to iodinated contrast, making this the safest of the current noninvasive modalities, except in those patients with pacemakers, implantatble defibrillators or recent stent placement. The new MRI techniques may also allow quantification of velocity and flow in coronary arteries, as well as assessment of plaque morphology. MR imaging has proved successful in producing angiograms of peripheral vascular anatomy and abnormalities. MRI has already been very successful in the detection of coronary artery variants and in the imaging of coronary stents and bypass grafts.

SUGGESTED READING

1. Popp RL, Fowles R, Coltart J. Cardiac anatomy viewed systematically with two-dimensional echocardiography. Chest 1979; 75:579.

2. Reeder GS, Seward JB, Tajik AJ. The role of two-dimensional echocardiography in coronary artery disease: A critical appraisal. Mayo Clin Proc 1982; 57:247.

3. Rubenson DS, Tucker CR, London E et al. Two dimensional echocardiographic analysis of segmental left ventricular wall motion before and after coronary artery bypass surgery. Circulation 1982; 66:1025.

4. Jones RH, McEwan P, Newman GE et al. Accuracy of diagnosis of coronary artery disease by radionuclide measurement of left ventricular function during rest and exercise. Circulation 1981; 64:586.

5. Iskandarian A, Hakki AH. Radionuclide evaluation of exercise left ventricular performance in patients with coronary artery disease. Am Heart J 1985; 110:851-856.

6. Bonow R, Keat K, Rosing D et al. Exercise-induced ischemia in mildly symptomatic patients with coronary artery disease and preserved left ventricular function. N Engl J Med 1984; 311:1339-1345.

7. Noto TJ Jr, Johnson LW, Krone R et al. Cardiac catheterization 1990: A report of the Registry of the Society for Cardiac Angiography and Interventions (SCA&I). Cathet Cardiovasc Diagn 1991; 24:75-83.

8. Achenbach S, Moshage W, Ropers D et al. Value of electron-beam computed tomography for the noninvasive detection of high-grade coronary artery stenoses and occlusions. N Engl J Med 1998; 339:1964-71.

9. Goldin JG, Yoon HC, Greaser LE III et al. Spiral versus electron-beam CT for coronary artery calcium scoring. Radiology 2001; 221:213-21.

10. Kim WY, Danias PG, Stuber M et al. Coronary magnetic resonance angiography for the detection of coronary stenoses. N Engl J Med 2001; 345:1863-9.

11. Kwok K, Kim C, Grady D et al. Meta-analysis of exercise testing to detect coronary artery disease in women. Am J Cardiol 1999; 83:660-6.

Cardiopulmonary Bypass and Myocardial Protection

Fritz J. Baumgartner

3

GENERAL PRINCIPLES

Cardiopulmonary bypass (Fig. 3.1) is a process by which systemic venous blood is taken from the patient, transferred to a pump oxygenator and delivered back to the arterial circulation of the patient. Cardiopulmonary bypass (CPB) can be performed either at normothermia or by cooling the patient to 30°C, 28°C or even as low as 15-18°C in preparation for complete circulatory arrest. Cardioplegia is induced by a high potassium-containing solution which causes electromechanical arrest of the myocardium. Electromechanical arrest decreases myocardial oxygen utilization substantially; it is said that there is a 90% decrease in myocardial oxygen utilization by simply keeping the heart at a standstill. An additional 10% decrease in myocardial oxygen consumption is achieved by cooling the myocardium. The bypass circuit at Harbor-UCLA consists of a single venous cannula in two stages for coronary bypass surgery in which inferior vena cava blood goes through the cannula and a proximal port in the right atrium drains the remainder of the systemic venous blood coming from the superior vena cava. In the case of mitral valve replacement or situations in which the right heart must be entered, the venous cannulation consists of a superior vena cava cannula and a separate inferior vena cava cannula. In the case of mitral valve replacement, this is useful because it keeps all of the hardware out of the field of the mitral valve and permits increased mobilization of the left atrium and increased visualization of the mitral valve. In the case of right heart surgery, as for example, tricuspid valve or atrioseptal defect surgery, this is important because without bicaval cannulation, once the right heart is entered air will be aspirated into the venous cannula and cause an air lock, shutting off the pump.

The venous line drains down to a reservoir. The amount of venous drainage can be regulated by the height of the reservoir; i.e., the higher the reservoir the less the venous drainage, and the lower the reservoir with respect to the patient the higher the amount of venous drainage (i.e., the effect of gravity). The reservoir blood then enters a hollow fiber membrane pump oxygenator (Fig. 3.1) with a temperature regulating device in the proximal portion of the system and the oxygenator just distal to this. The entire heater/warmer oxygenator circuit is just distal to the roller pump head. (In the case of bubble oxygenators, the oxygenator is

Cardiothoracic Surgery, Third Edition, edited by Fritz J. Baumgartner. ©2004 Landes Bioscience.

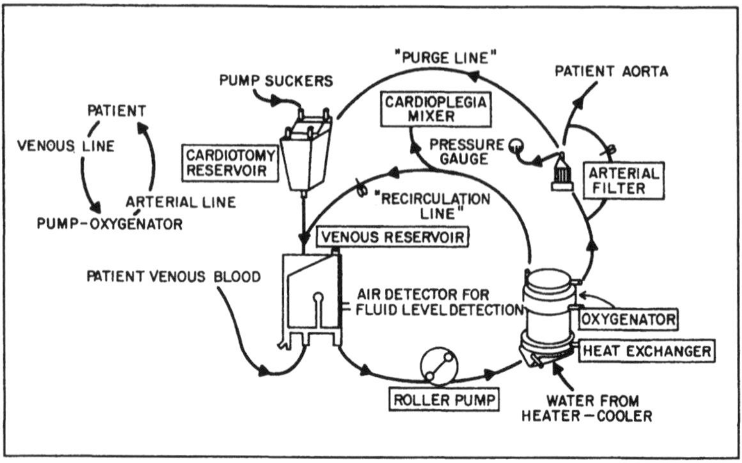

Fig. 3.1. Cardiopulmonary bypass circuit.

proximal to the roller pump head.) Once the blood passes through the membrane where the CO_2/O_2 exchange takes place, the blood travels through a 40 micron filter and then back into the arterial circuit of the patient. The filter serves to remove particulate and gaseous emboli. The arterial circuit has a purge line which can remove gross air.

Cooling in cardiopulmonary bypass is done at approximately 1°C per minute. The danger of rapid cooling is that if the cool blood enters the warm patient, gaseous emboli can form within the patient, but this should not happen if the gradient between the infused blood and the patient is less than 10°C. The advantages of cooling are that it decreases the metabolic requirement of the body organs, in particular, the brain and the heart. However, disadvantages are that it may increase bleeding after coming off bypass because of stunning of the coagulation enzyme systems, and it may induce myocardial edema by impairment of enzyme systems.

Circulatory arrest is another modality which is used in tiny infants who require cardiopulmonary bypass and then removal of their cannulas after achieving circulatory arrest to prevent clutter of the operative field. It is also used in specific situations: transverse aortic arch aneurysms, inferior vena cava surgery, or descending thoracic/thoracoabdominal aortic surgery where placement of an aortic cross-clamp or venous cross-clamp is awkward, and there is a need to arrest the circulation to keep the surgical field clear of blood to permit suturing.

CARDIOPLEGIA

Cardioplegia solution is used to arrest and protect the heart. The solution used at Harbor consists of 30 or 60 mEq/liter K^+ D5-1/4 normal saline containing solution buffered with Tham and CPD. The solution is diluted in a 1:4 mixture with blood drained from the patient. Thus, the final concentration of potassium in the low and high potassium containing solutions is in the order of 10 and 20 mEq/liter respectively. High potassium containing cardioplegia is used for the initial arrest. This is then switched to low potassium cardioplegia to minimize the amount of total potassium load that the patient receives to prevent hyperkalemia and arrhythmias while attempting to wean from cardiopulmonary bypass. In general, at Harbor cardiac arrest and myocardial protection is achieved with both antegrade and retrograde cold blood cardioplegia. There are several factors to consider regarding cardioplegia: (1) cold versus warm; (2) antegrade versus retrograde; (3) blood versus crystalloid. In general, we prefer cold blood antegrade cardioplegia at 4°C. There are many advocates of continuous warm blood cardioplegia. It prevents damage to enzyme systems resulting from cold temperatures, and thus the sodium potassium ATPase is kept intact, helping to limit the amount of cellular swelling and third space edema within the myocardium. Also it prevents damage to the coagulation enzymes, thus reducing bleeding once cardiopulmonary bypass is terminated. Also it appears that weaning from cardiopulmonary bypass is easier with warm cardioplegia, and the patients more often come off in normal sinus rhythm. It should be noted that 90% of oxygen utilization by the myocardium is eliminated simply by arresting the heart, which warm cardioplegia does quite efficiently; only another 10% of myocardial protection is added by cooling the myocardium. It should be noted, however, that advocates of continuous warm blood cardioplegia still maintain that in any condition where there is a question of myocardial protection, standard cold cardioplegia should be immediately instituted, since this is a time-honored method of protecting the heart. At Harbor we tend to use cold blood cardioplegia since it can be given intermittently rather than continuously and affords a small additional decrease in myocardial oxygen consumption. One additional problem with warm continuous cardioplegia is that it obscures the surgical field due to blood. Another is safety. If a problem arises with the pump while warm, the margin of safety is greatly diminished.

Blood cardioplegia is felt to be superior to crystalloid cardioplegia because blood delivers some amount of oxygen to the myocardium, even though the myocardial oxygen utilization is at a low rate. Nonetheless, the oxygen utilization of the myocardium does exist and blood facilitates its delivery.

Retrograde cardioplegia is felt to be superior to the antegrade method. In general, we start off with an antegrade dose to arrest the heart and then switch to retrograde soon thereafter. The retrograde system is particularly useful for redo coronary surgery and eliminates the possibility of atheromatous embolization of debris from old grafts ("trash heart"). It also provides more uniform cooling of the myocardium as well as better distribution of cardioplegia solution beyond blockages in the coronary arteries. Finally, retrograde cardioplegia administration has the advantage that the cardioplegia may be given continuously and the heart may be manipulated and moved around without concern for aortic regurgitation when administered through the aortic root.

TECHNIQUE

In summary, cannulation and cardiopulmonary bypass is instituted by placing aortic pursestrings in the distal ascending aorta followed by venous pursestrings. The venous pursestring is either a single right atrial pursestring in the case of coronary bypass surgery or aortic valve replacement, or in the case of mitral valve replacement or tricuspid valve procedures, a separate superior vena cava and inferior vena cava pursestring. Figure 3.2 shows the cardiac cannulation sites for various procedures as well as the different types of cannulas. After the aortic pursestring has been applied, the patient is heparinized. After the venous pursestring has been placed, the antegrade cardioplegia pursestring is applied followed by the retrograde cardioplegia pursestring. The aorta is cannulated followed by venous cannulation of either the right atrium alone or the superior and inferior vena cava separately. Then the antegrade plegia line is inserted into the aorta. The retrograde plegia line is inserted through the right atrium directly into the coronary sinus blindly by simply palpating the back of the heart and feeling where the retrograde cannula enters the coronary sinus. Cardiopulmonary bypass is instituted when the activated clotting time is above 400 seconds. In tricuspid valve procedures or anytime the right heart is entered, it is necessary to place caval tapes around the superior and inferior vena cava so that when the right heart is entered air will not be aspirated into the venous cannulae forming an air lock. The tapes

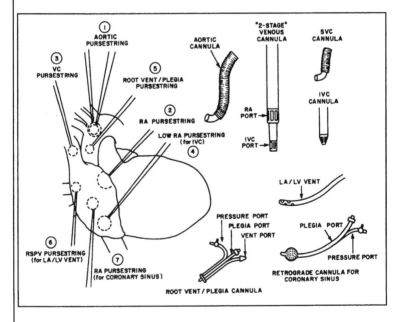

Fig. 3.2. Cannulas and cardiac cannulation sites. CABG: 1, 2, 5, 7; AVR: 1, 2, 5, 6, 7; MVR: 1, 3, 4, 5, 6, 7; TVR: 1, 3, 4, 5 . . . and caval snares.

prevent blood from rushing up into the operative field and causing impaired vision.

After cardiopulmonary bypass is instituted, a pursestring is placed in the right superior pulmonary vein for cases of aortic and mitral valve procedures. The different types of cannulas are shown in Figure 3.2. The aortic cannula is shown, as well as the two-stage venous cannula for coronary bypass surgery or for aortic valve replacement. The first stage lies within the right atrium and the second stage lies within the inferior vena cava. A separate superior vena cava and inferior vena cava cannula are shown. The antegrade cardioplegia cannula is shown and actually serves three functions in one. One is to administer cardioplegia; the second is to monitor pressure through a separate line; and the third is to serve as a vent for the aortic root through a separate port. The retrograde cannula for insertion into the coronary sinuses is shown as well and this too has a pressure port as well as a cardioplegia port. The pressure port is very important for the retrograde cannula because overpressurizing the coronary sinus may result in rupture. Pressures should be kept in the range of 20-40 mmHg.

The left atrial/left ventricular vent is shown. The purpose of the vent is 3-fold. First, it prevents overdistension of the heart which would impede coronary perfusion and lead to a very dangerous situation. Second, it evacuates blood which obscures the surgical field. Third, it evacuates air from the left heart during de-airing maneuvers. In the case of aortic valve replacement, this vent actually goes into the left ventricle through the right superior pulmonary vein and left atrium. This differs from mitral valve replacement in which the vent tip lies initially in the left atrium. Only after the mitral procedure has been performed is the left atrial vent pushed visually into the left ventricle.

Methods of venting the left ventricle include the standard vent going through the right superior pulmonary vein through the left atrium, into the left ventricle. Another useful technique involves directly stabbing the apex of the left ventricle and inserting a vent directly through it; this is useful as an emergency procedure. The other way of venting the left ventricle is to place a vent in the pulmonary artery which then aspirates blood retrograde from left heart through the pulmonary veins.

If the heart suddenly distends for no apparent reasons while on bypass and there is no left ventricular vent, one needs to decompress the heart immediately. A distended left ventricle results in inadequate perfusion of the myocardium. One can also decompress the heart by shutting off the pump allowing the venous blood to drain for a few seconds and allowing blood to go from the left ventricle up through the aortic root vent. One can then reduce flow through the aortic line to keep the heart decompressed. If on cardiopulmonary bypass there is an extremely low perfusion pressure in the 20-30 mmHg range despite being on full flow, then one must suspect the possibility of aortic insufficiency and a vent should be placed if not already present via the right superior pulmonary vein into the left atrium and into the left ventricle. If this still does not help and there is still a low perfusion pressure while on cardiopulmonary bypass, the aortic cross-clamp should be applied to prevent regurgitation and the aortic root will then have to be opened in order to administer antegrade cardioplegia directly into the coronary orifices.

Fig. 3.3. Femoral-femoral
bypass.

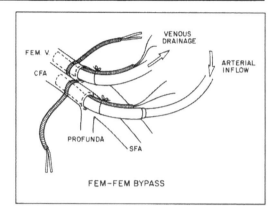

FEM-FEM BYPASS

To summarize, in coronary artery bypass surgery, one requires an aortic cannula, a two-stage venous cannula, an antegrade cardioplegia cannula, and a retrograde cardioplegia cannula. For aortic valve replacement, one requires an aortic cannula, a two-stage venous cannula, an antegrade plegia cannula, a retrograde plegia cannula, and a left ventricular vent. For mitral valve replacement one requires an aortic cannula, a separate superior vena cava and inferior vena cava cannula, an antegrade plegia cannula, a retrograde plegia cannula, and a left atrial vent. For tricuspid valve replacement, one requires an aortic cannula, separate superior and inferior vena cannulae, antegrade plegia cannula, and caval tapes about the superior and inferior cavae.

FEMORAL-FEMORAL BYPASS

This is a maneuver by which full cardiopulmonary bypass can be established through the femoral artery and femoral vein (Fig. 3.3). Venous blood is taken from the femoral vein, transferred to the pump oxygenator and returned through the femoral artery to perfuse the patient's entire body. Blood from the femoral artery goes retrograde up the thoracic aorta to perfuse the aortic arch vessels and the coronary arteries. There are several instances in which femoral-femoral bypass may be required. One is a patient who had previous open heart surgery and is an extremely high risk for cardiac arrest during dissection of the previous adhesions. In such a situation, it may be advisable to prepare for femoral-femoral bypass since the aorta and right atrium may not be easily surgically accessible in the event of a cardiac arrest. Another situation where femoral-femoral bypass may be helpful is in descending thoracic or thoracoabdominal operations in which spinal cord protection may be afforded below the level of the cross-clamp by using femoral-femoral bypass. This technique may also be helpful for rewarming patients who present with hypothermic cardiac arrest, allowing rapid rewarming while maintaining cardiopulmonary support.

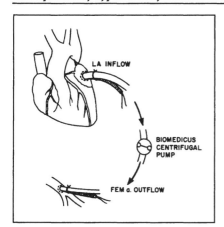

Fig. 3.4. Left atrial-femoral bypass.

3

LEFT ATRIAL-FEMORAL BYPASS

Left atrial femoral bypass is partial bypass in which there is no oxygenator involved. Left atrial flow is transmitted through a centrifugal pump to the femoral artery (Fig. 3.4). It is used in conditions where there is a need for spinal protection below the level of the cross-clamp, and this may include descending thoracic and thoracoabdominal operations in a similar way that femoral-femoral bypass is utilized.

INTRA-AORTIC BALLOON PUMP

The intra-aortic balloon pump (IABP) is inserted via the femoral artery and lies within the descending thoracic aorta. It increases coronary perfusion by inflating during diastole and decreases afterload against which the heart must eject by deflating just before systole. It is by these two mechanisms that the IABP improves cardiac performance. It is most often used intraoperatively to support a failing ventricle (i.e., a ventricle that will not come off cardiopulmonary bypass). Other important indications include: (1) intractable angina pectoris; (2) cardiogenic shock or low output state outside of the operating room; (3) mitral regurgitation or acute ventricular septal defect secondary to infarctions; and (4) uncontrolled malignant arrhythmias.

Contraindications to an IABP include severe aorto-iliac disease that precludes entering the femoral artery. If an IABP is absolutely required despite this contraindication, it may be necessary to place the balloon into the ascending aorta and thread it into the descending thoracic aorta. Other contraindications include aortic insufficiency which would worsen with inflation of the balloon. Another contraindication is aortic dissection because of possible injury to the aorta by the balloon, although it has succeeded in selected instances.

PATHOPHYSIOLOGY OF CARDIOPULMONARY BYPASS

Cardiopulmonary bypass is not physiologic. It results in several metabolic derangements including a bleeding tendency (secondary to platelet sequestration, clumping in the oxygenator, and diffuse systemic inflammatory reaction with third spacing); this will be discussed in detail.

Vasoconstriction occurs as a result of the catecholamine storm, hemolysis secondary to red blood cell trauma and sheer stress and protein denaturation within the oxygenator. The diffuse systemic inflammatory reaction occurring from cardiopulmonary bypass all centers about activation of Hageman factor 12 by the cardiopulmonary bypass foreign circuit. Hageman factor 12 activates four pathways: (1) fibrinolysis; (2) the clotting cascade such that even in the presence of heparin consumption of coagulation factors occurs to some degree; (3) the complement cascade including C3A and C5A resulting in increased vascular permeability, third spacing, and neutrophil migration; (4) the kallikrein/bradykinin system resulting in increased vascular permeability and pain.

Cardiopulmonary bypass is a stress on the body similar to the stress of trauma or surgery itself. The concentration of epinephrine in the plasma increases 10-fold in cardiopulmonary bypass and that of norepinephrine increases three to four times. ADH during cardiopulmonary bypass far exceeds that of other surgical procedures. There is increased lipolysis from the increased epinephrine and hyperglycemia with impaired insulin responsiveness in cardiopulmonary bypass.

Bypass causes a 7-fold increase in thyroxin levels secondary to the heparinization. ACTH levels fall during cardiopulmonary bypass with resulting decrease in cortisol. Pulsatile cardiopulmonary bypass minimizes the fall of ACTH, and thus the ACTH and cortisol are closer to normal than during continuous flow perfusion.

Bleeding is common when coming off cardiopulmonary bypass because of sequestration of platelets, inactivation of coagulation factors from enzymatic injury, as well as stunning of the enzyme systems from hypothermia. Many techniques to achieve hemostasis in the operating room are utilized. Blood products including fresh frozen plasma, cryoprecipitate, and platelets may be utilized. Amicar (alpha-aminocaproic acid) is an antifibrinolytic and prevents clot destruction. DDAVP, Gelfoam and thrombin, and "cryoglue" (a combination of thrombin, calcium, and cryoprecipitate which coagulates when mixed together) may be used.

Special reference should be made to aprotinin (Trasylol) an anti-inflammatory, antifibrinolytic agent which appears to have a profoundly beneficial action in limiting coagulopathy-related bleeding. Aprotinin is a protease inhibitor with a variety of effects on the coagulation system. It blocks fibrinolysis by inhibiting plasmin and kallikrein. It inhibits the contact phase of coagulation and prevents consumption of coagulation factors. Aprotinin also preserves glycoproteins in the platelet membrane making them resistant to damage from cardiopulmonary bypass. It is used prophylactically in patients at high risk for bleeding, including reoperation, aortic dissections and endocarditis.

Protamine is given after completion of cardiopulmonary bypass to reverse the heparin. A protamine reaction results in decreased systemic vascular resistance with hypotension and sometimes pulmonary vasocontriction that may lead to right heart failure. It may be related to a direct effect and/or an autoimmune or anaphylactic reaction. It also activates complement by the classical pathway. The treatment of a full-blown protamine reaction is to administer isoproterenol, epinephrine or prostaglandin E_1 into the right heart and calcium into the left heart. Intravenous Amrinone which acts as a cardiac inotrope and pulmonary vasodilator may be useful as well. Patients previously receiving the NPH form of insulin, containing protamine, may be presensitized and particularly susceptible to a protamine reaction.

The ACT (activated clotting time) for safe cardiopulmonary bypass without risk of clotting the oxygenator should be greater than 400 seconds. The ACT is inaccurate for high heparin doses or in hypothermia. The ACT may be prolonged with inadequate levels of clotting factors even if heparin concentrations are low.

Generally, during systemic cooling on bypass no greater than 10°C gradient (8°C for children) should exist between the heat exchanger and the venous blood. When cooling, the pO_2 of the arterial blood should be less than 200 mmHg lest bubbles form when the cool blood enters warm tissue. Rewarming tends to be slower than cooling because the risk of air embolization is greatest on warming. The heat exchanger temperature should not be greater than 40°C. Cooling is generally done at a rate of 1°C per minute; rewarming is done at 1°C every 3-5 minutes.

Generally, it has been stated that for each hour of cross-clamp time, one needs approximately 10 minutes of reperfusion (supportive bypass). This is often not the case and weaning may frequently be done quite rapidly. Flow in cardiopulmonary bypass should be 2.4 liters/min/m² at normothermia. This flow can be less with hypothermia. For each decrease in temperature of 1°C oxygen consumption by tissue decreases by 10%. Between 22-37°C, cerebral blood flow is not altered, but when the temperature becomes less than 22°C, cerebral blood flow is markedly less and is in the range of 15% of normothermic blood flow.

AUTOLOGOUS TRANSFUSION

Some patients are ideal candidates for autologous blood transfusion since they have increased central blood volume; these include patients with valvular disease and congestive heart failure, as well as patients with intracardiac shunts with cyanosis. The classic example is the polycythemia associated with tetralogy of Fallot. It is especially useful in patients with rare blood types or in those with circulating antibodies making crossmatch difficult. Drawing blood for use in autologous blood transfusions can be done 3-4 weeks preoperatively. Pre-bypass removal of one to two units of blood from the patient is also useful. This can be done through the CVP, aortic or venous cannulae.

pH AND CO_2 MONITORING ON CPB

There are several methods of assessing blood pH and carbon dioxide levels. One must first understand the normal physiologic change of carbon dioxide concentrations and blood pH changes with the temperature. At 37°C, the pH is 7.4 whereas in the periphery where the temperature is 25°C, the pH is 7.6. A formula to describe this is that the change in pH for every degree Celsius is -0.017, i.e., for every degree Celsius the patient drops, the pH goes up by 0.017. This temperature-pH slope is similar to that of water's neutral pH and results from a constant OH^- concentration with regard to to H^+ concentration over a great spectrum of temperatures. This is done to preserve the charge state of the dominant protein buffer imidazole. Thus, when progressing from high temperature to low temperature, to maintain pH the body compensates by decreasing the CO_2. Thus at lower temperatures organisms have a higher pH and lower CO_2. One must keep this in mind when measuring pHs while on cardiopulmonary bypass. The *alpha stat* method of blood pH maintenance on cardiopulmonary bypass is such that the carbon dioxide and pH are not corrected for temperature; i.e., the increased pH and decreased CO_2 at hypothermic situations is not corrected. With the *pH stat* method of blood pH maintenance on cardiopulmonary bypass, the CO_2 and pH are corrected, i.e., the increased pH and decreased CO_2 at hypothermic conditions are corrected by infusing more CO_2 into the cardiopulmonary bypass circuit, thus raising the patient's CO_2 and lowering the pH. Most authorities now agree that the pH stat method is less physiologic than the alpha stat method. Some say, however, that the pH stat method of infusing CO_2 will result in better cerebral blood perfusion because of cerebral vasodilation, but this has not been shown to be clinically important.

AIR EMBOLUS

"Never open a beating heart" is an axiom referring to the left heart when the aorta is unclamped (i.e., one may still open the right atrium for repair of an ASD or tricuspid valve repair since this is on the right side of the heart). Opening the left atrium is much more dire than opening the left ventricle because of the vacuum effect of the atrium sucking in air. The most common cause of air embolism is inattention by the perfusionist to the oxygenator blood level. If the blood level falls below a certain critical level air will be sucked into the pump. Fortunately, the pumps now in use will automatically shut off when air is sensed. Another cause is an unexpected resumption of heartbeat while the heart is open and the cross-clamp is off. Reversal of left ventricular vent suction, such as air pumped directly into the left ventricle, is another cause. Opening a low pressure right superior pulmonary vein while off bypass, with the cross-clamp off and the heart beating may result in air being aspirated into the left heart. Detachment of the oxygenator during perfusion may result in air embolus. Faulty technique during circulatory arrest, in which air is left in the aorta, may result in air embolus.

In case gross air embolism occurs, the following emergency maneuvers are undertaken. The pump is immediately shut off and a clamp is placed on the arterial line. The arterial line is disconnected from the arterial cannula and placed onto a cannula in the superior vena cava. The patient, meanwhile, is placed in Trendelenburg as the anesthesiologist massages the carotid arteries retrograde. Flow through the superior vena cave is instituted at 2 l/min for 2 minutes and blood is retrogradely flushed through the superior vena cava through the brain, down the carotid arteries and back out through the arterial cannula into the pericardial sac which is then sucked up with the pump sucker to be returned to the pump circuit. Mannitol and pentobarbital should be given. If an air embolus does occur, the patient should be cooled to afford some cerebral protection. The patient should also be ventilated with 100% FIO_2 to facilitate the resorption of nitrogen. Hyperbaric oxygen can be used postoperatively to increase oxygen delivery and increase solubility of gas emboli.

The largest source of fragment microembolization in cardiopulmonary bypass is field aspirated pump sucker blood, since this causes trauma of the blood components and sends back unwanted debris. Arterial filters are therefore used in the cardiopulmonary bypass circuit. Arterial filters usually have a pore size of 40 microns and filter gaseous as well as particulate emboli from the blood before it is returned to the patient.

CARDIOPULMONARY BYPASS DURING PREGNANCY

This is usually done for rheumatic mitral valve disease or endocarditis. Several caveats hold: (1) the procedure should be done after the first trimester but before the third trimester in order to avoid the period of organogenesis and the period where the cardiac demands and the blood volume are maximal; (2) keep the fetus adequately oxygenated and avoid acidosis; (3) keep the patient tilted slightly to the left to permit adequate venous return from the inferior vena cava as blood flow is 30-50% more in pregnancy; (4) the perfusion pressure should be kept at least 60 mmHg; (5) a fetal heart monitor should be in place; (6) the patient should not be cooled further than approximately 32-34°C to ensure that the fetus is not at risk of fibrillation due to low temperatures.

SICKLE CELL ANEMIA

Patients with sickle cell anemia are more fragile on cardiopulmonary bypass. It is crucial to keep the oxygen and pH adequate. Acidosis shifts the oxygen/hemoglobin dissociation curve to the right, thus unloading the red blood cell of oxygen resulting in sickling of the red blood cell. If sickling occurs intraoperatively on cardiopulmonary bypass, increased mechanical fragility of the walls and trauma of bypass may increase hemolysis. Some sickling is reversible by simply increasing oxygenation. Even the sickle cell trait may result in sickling and cause organic infarct under cardiopulmonary bypass.

CARDIOPULMONARY BYPASS IN CHRONIC RENAL DISEASE

Cardiopulmonary bypass in chronic renal disease is usually done for procedures involving coronary bypass or valve replacement. Myocardial infarction is the cause of death in one-third of patients on hemodialysis. Chronic anemia in the presence of the AV fistula used for dialysis contributes to ischemic changes. Preoperative cardiac catheterization can be life-threatening in these patients from the fluid shifts and potassium arrhythmias, hence the patient should be dialyzed in the morning prior to catheterization and again the evening prior to surgery. One could consider putting in femoral instead of radial arterial lines in anticipation of future AV fistulas. The perfusion flow should be the standard 2.4 l/min/m². Hourly potassium levels are checked and levels greater than 6.0 mEq/l are treated with glucose and insulin. One can perform hemodialysis following, or even during, cardiopulmonary bypass. One should at least use an ultrafiltrator to remove volume accumulated during cardiopulmonary bypass that cannot be cleared by the kidney.

CNS INJURY IN SURGERY OF THE DESCENDING AORTA

For some patients, up to 45 minutes of descending aortic occlusion is well tolerated, while in others paraplegia occurs after only 10 minutes. The discrepancy may be related to variability in the blood supply to the spinal cord. There are three major regions of blood supply: (1) cervical cord is supplied from the vertebral branch of the subclavian artery via the anterior spinal artery; (2) the mid-thoracic cord is supplied by spinal arteries rising from the level of about T7; (3) the lower cord is supplied from the large unpaired artery of Adamkiewitz arising in 90% of patients from T9 to L2 and in about 10% from T5 to T8. This contributes to the anterior spinal artery which can provide collateral circulation throughout the spinal cord. The anterior spinal artery can be continuous along the entire spinal cord or can be discontinuous. Anterior spinal occlusion causes paraplegia, incontinence, and loss of superficial sensation. Deep pin, vibration and position sensation, transmitted via the posterior columns, are preserved. If the anterior spinal artery is discontinuous, proximal occlusion can create ischemia and paraplegia. Intraoperative methods to determine spinal viability include somatosensory evoked potentials and somatomotor evoked potentials. Some surgeons find them of limited use, however.

During cross-clamping of the descending aorta, extravascular conduits may be extremely helpful to protect the spinal cord. This includes left atrial-femoral arterial bypass, femoral-femoral bypass, or a passive shunt (Gott shunt from the proximal aorta to the distal aorta or to the femoral artery). In the case of aortic transection, however, none of these methods have been shown to definitively decrease the incidence of paraplegia.

Most often, however, if the cross-clamp time can be limited to less than 20 minutes, the "clamp and go" method is used without specific spinal protection methods other than speed of surgery. If the cross-clamp time exceeds 20-30 minutes, the incidence of paraplegia rises dramatically.

INTRAOPERATIVE HEART FAILURE

In the operating room, to differentiate left heart failure from right heart failure, one may be able to get some idea simply by looking at the heart. If the right ventricle appears to be sluggish and the left ventricle is contracting normally, the answer may be clearly visible and vice versa. Hemodynamic parameters can also give us some idea whether there is right heart failure versus left heart failure. If there are high pulmonary artery pressures and high CVP, (i.e., elevated right heart pressures), and despite this the left ventricular wedge pressure is low and cardiac output is poor, then right ventricular failure is presumed to have occurred. Conversely, if an elevated wedge pressure exists but the cardiac index is low then one can presume left heart failure. This will help determine whether an RVAD, LVAD or both are necessary for weaning from cardiopulmonary bypass. If the surgeon is unable to re-animate the heart after cardiopulmonary bypass, the problem may be either distention of the heart, electrolyte or blood gas problems, or the problem may be automaticity or conduction problems which can be solved by pacing. If the patient is severely hyperkalemic and has related myocardial dysfunction and the standard methods of lowering the potassium are ineffective, (i.e., glucose, insulin, kayexelate, bicarbonate, calcium and lasix), then hemodialysis is necessary. If the patient has cardiac asystole refractory to pacing, but if inspection of the heart reveals good color without massive edema or hemorrhage, the prognosis is good and one may be able to convert into ventricular fibrillation using calcium or epinephrine. If there is terminal asystole, such as a heart severely damaged by anoxia, poor myocardial protection, etc., this is ominous and is best managed by a period of cardiopulmonary bypass to let the heart rest. A case of sustained ventricular contraction, i.e., stone heart (synonymous with a calcium paradox) can sometimes be treated with Inderal or reperfusion with cardioplegia. The cause is often the absence of calcium within the cardioplegia solution which can be corrected by reinfusing the heart with calcium-containing blood. A small amount of calcium must always be used in the cardioplegia solutions to prevent this phenomenon from occurring. For ventricular fibrillation refractory to shocking, it must be confirmed that electrolytes and blood gases have normalized and that anti-arrhythmics have been given. If the patient is still refractory, one should cross-clamp the aorta again and give 500 cc of cardioplegia at body temperature, then release the cross-clamp and pace the heart. If the heart continues to degenerate to ventricular fibrillation despite a brief period of normal contraction, one may try to pace the heart during a normal period to prevent fibrillation.

Areas of subendocardial ischemia are better treated by nitroglycerin than Nipride. Nitroglycerin increases the endocardial to epicardial blood flow ratio. It is a less powerful arterial dilator than Nipride and so maintains myocardial perfusion pressure. Nipride increases the epicardial to endocardial flow ratio. It is a

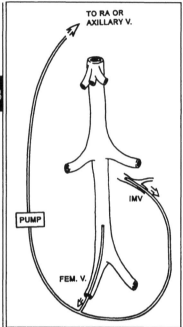

TO RA OR
AXILLARY V.

IMV

PUMP

FEM. V.

Fig. 3.5. Schematic diagram of VVB. Systemic blood is actively drained from the femoral vein via a centrifugal pump to the central venous system. Portal decompression may be achieved via the inferior mesenteric vein.

powerful arterial dilator and the resulting steal syndrome would flow away from fixed obstructions. The catecholamine levels fall to normal within 20 minutes of cardiopulmonary bypass and by 4-8 hours postoperatively the sympathetic tone decreases and the extremities begin to vasodilate accounting in part for the myocardial dysfunctions seen 4-8 hours postoperatively.

VENOVENOUS BYPASS

Another extracorporeal technique used in certain special situations is venovenous bypass (VVB). It is used predominantly for vascular isolation of difficult inferior vena cava procedures. VVB was originally described for the anhepatic phase of liver transplantation. Other uses include the resection of renal and adrenal tumors with caval involvement and trauma to the retrohepatic inferior vena cava. VVB drains blood from the lower half of the body via a centrifugal pump into the central venous circulation. The systemic inflow into the circuit is from the femoral vein and the outflow is either the right atrium or axillary vein (Fig. 3.5). The portal system can be drained with a separate cannula into the inferior mesenteric vein. Flows are in the range of 2-3 l/min and because there is no oxygenator in the circuit, the blood remains deoxygenated. Heparinization is usually in the range of 100 units/kg, although cases have been successfully done without systemic heparin.

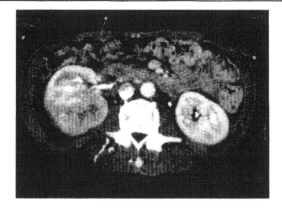

Fig. 3.6a. CT scan of a right renal cell carcinoma with tumor extending up the inferior vena cava.

Clamping the inferior vena cava leads to rapid central hypovolemic shock because of sequestration of blood in the lower extremities without return to the heart. VVB allows for vascular isolation of the inferior vena cava with continued return of lower extremity venous blood to the heart. Drainage of the portal system into the VVB circuit prevents the accumulation of toxic gut metabolites which may be arrhythmogenic when the cross-clamps are released. The complete VVB circuit is shown in Figure 3.6 for resection of a right renal cell carcinoma involv-

Fig. 3.6b. The venovenous bypass circuit drains femoral venous and portal blood into the right atrium. Vascular control of the infrarenal cava, intrapericardial cava, right renal artery, left renal vein, hepatic artery and portal vein is shown.

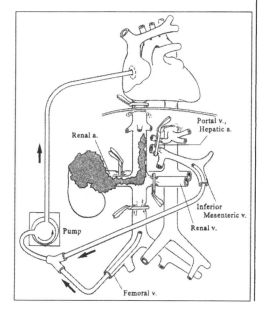

Fig. 3.7a. Right renal cell carcinoma growing up the inferior vena cava into the right atrium. Cardiopulmonary bypass, profound hypothermia and circulatory arrest are required for resection of such a tumor.

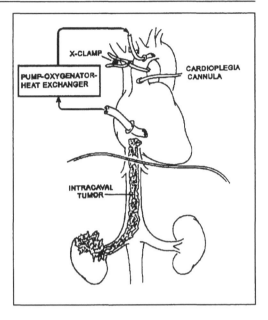

Fig. 3.7b. Example of a renal cell carcinoma resected with this technique.

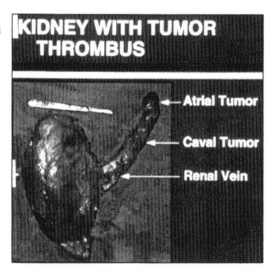

ing the cava. It is important to recognize that if the tumor thrombus extends up into the right atrium, VVB is no longer technically possible because snare control of the intrapericardial inferior vena cava will trap and break off the tumor. To resect such a tumor, it may be necessary to establish full cardiopulmonary bypass, cool to profound hypothermia, and establish circulatory arrest. Figure 3.7 depicts such a tumor requiring circulatory arrest for resection.

SUGGESTED READING

1. Gibbon JH Jr. Application of a mechanical heart and lung apparatus in cardiac surgery. Minn Med 1954; 37:171.
2. Lillehei CW. A personalized history of extracorporeal circulation. Trans Am Soc Artif Intern Organs 1982; 28:5-16.
3. Brunberg JA, Reilly EL, Doty DB. Central nervous system consequences in infants of cardiac surgery using deep hypothermia and circulatory arrest. Circulation 1974; 49-50(Suppl 2):60.
4. Gay WA Jr, Ebert PA. Functional metabolic effects of potassium-induced cardioplegia. Surgery 1976; 54:193.
5. Robertson JM, Vinten-Johansen J, Buckberg GD et al. Safety of prolonged aortic clamping with Blood Cardioplegia. J Thorac Cardiovasc Surg 1984; 88:395-401.
6. Arom KV, Vinas JF, Fewel JE et al. Is a left ventricular vent necessary during cardiopulmonary bypass? Ann Thorac Surg 1977; 24:566.
7. Hammerschmidt DE, Strancek DS, Bowers TK et al. Complement Activation and Neutropenia occurring during Cardiopulmonary Bypass. J Thorac Cardiovasc Surg 1981; 81:370.
8. Chenoweth DE. Complement activation during cardiopulmonary bypass. In: Utley JR, ed. Pathophysiology and Techniques of Cardiopulmonary Bypass. Vol 2. Baltimore: Williams & Wilkins, 1983:49-60.
9. Wareing TH, Davila-Roman VG, Daily BB et al. Strategy for the reduction of stroke incidence in cardiac surgical patients. Ann Thorac Surg 1993; 55:1400-8.
10. Downing SW, Savage EB, Streicher JS et al. The stretched ventricle: myocardial creep and contractile dysfunction after acute nonischemic ventricular distension. J Thorac Cardiovasc Surg 1992; 104:996-1005.
11. Kestin AS, Valerie CR, Khurie SF et al. The platelet function defect of cardiopulmonary bypass. Blood 1993; 82:107-17.
12. Kurusz M, Butler BD. Embolic events and cardiopulmonary bypass. In: Gravlee GP, Davies RF, Utley JR, eds. Cardiopulmonary Bypass. Baltimore:Williams & Wilkins, 1993:267-90.
13. Tabuchi N, de Haan J, Boonstra PW et al. Activation of fibrinolysis in the pericardial cavity during cardiopulmonary bypass. J Thorac Cardiovasc Surg 1993; 106:828-33.
14. Steinberg JB, Kapedanski DP, Olson JD et al. Cytokine and complement levels in patients undergoing cardiopulmonary bypass. J Thorac Cardiovasc Surg 1993; 106:1008-16.
15. Ko W, Hawes AS, Lazenby WD et al. Myocardial reperfusion injury. J Thorac Cardiovasc Surg 1991; 102:297-308.
16. Blauhut B, Klima U, Bettelheim P et al. Comparison of the effects of aprotinin and tranexamic acid on blood loss and related variables following cardiopulmonary bypass. J Thorac Cardiovasc Surg 1994; 108:1083-91.

17. Lu H, Saeria C, Commin P-L. Hemostasis in patients undergoing extracorporeal circulation: The effect of aprotinin (Trasylol). Thromb Haemost 1991; 66:633-7.

18. Hunt BJ, Segal HC, Jacoub M. Guidelines for maintaining heparin by the activated clotting time when aprotinin is used during cardiopulmonary bypass. J Thorac Cardiovasc Surg 1992; 103:211-2.

19. Gay WA Jr, Ebert PA. Functional metabolic effects of potassium-induced cardioplegia. Surgery 1976; 54:193.

20. Daggett WM, Randolph JD, Jacobs M et al. The superiority of cold oxygenated dilute blood cardioplegia. Ann Thorac Surg 1987; 43:397-402.

21. Salerno TA, Houck JP, Barrozo CAM et al. Retrograde continuous warm blood cardioplegia: A new concept in myocardial protection. Ann Thorac Surg 1991; 51:245-7.

22. Catinella FP, Cunningham JN, Adams PX. Myocardial protection with cold blood potassium cardioplegia during prolonged aortic cross-clamping. Ann Thorac Surg 1982; 33:228-33.

23. Ihnken K, Morita K, Buckberg GD et al. The safety of simultaneous arterial and coronary sinus perfusion: experimental background and initial clinical results. J Card Surg 1994; 9:15-25.

24. Loop FD, Higgins TL, Panda R et al. Myocardial protection during cardiac operations. J Thorac Cardiovasc Surg 1992; 104:608-18.

25. Allen BS, Hrat RS, Wiewall J et al. Retrograde cardioplegia does not adequately perfuse the rigth ventricle. J Thorac Cardivasc Surg 1995; 109:1116-26.

26. Matsuura H, Lazar HL, Yang X et al. Warm vs. cold blood cardioplegia: Is there a difference? Surg Forum 1991; 42:231-2.

27. Park SB, Liebler GA, Burkholder JA et al. Mechanical Support of the Failing Heart. Ann Thorac Surg 1986; 42:627-631.

28. Pae WE Jr, Miller CA, Matthews Y et al. Ventricular assist devices for postcardiotomy cardiogenic shock. J Thorac Cardiovasc Surg 1992; 104:541-3.

29. Bavaria JE, Furukawa S, Kreiner G et al. Effect of circulatory assist devices on stunned myocardium. Ann Thorac Surg 1990; 49:123-8.

30. Brunberg JA, Reilly EL, Doty DB. Central nervous system consequences in infants after cardiac surgery using deep hypothermia and circulatory arrest. Circulation 1974; 49-50 (Suppl 2):60.

31. Seelye ER, Harris EA, Squire AW et al. Metabolic effects of deep hypothermia and circulatory arrest in infants during cardiac surgery. Br J Anaesth 1971; 43:449.

32. Burt M. Inferior vena caval involvement by renal cell carcinoma. Use of veno-venous bypass as an adjunct during resection. Urol Clin North Am 1991; 18:437-444.

33. Baumgartner FJ, Scott R, Zane R et al. Modified venovenous technique for resection of renal and adrenal carcinomas with involvement of the inferior vena cava. Eur J Surg 1996; 162:58-62.

Cardiothoracic Anesthesia

John R. Charney

Cardiothoracic anesthesia has made great strides in just a few short years. Recent developments in the areas of monitoring devices and anesthetic agents and techniques have revolutionized our approach to the management of patients undergoing open heart and thoracic surgery.

MONITORING

Invasive monitoring of systemic and pulmonary artery pressures as well as cardiac output remain standard for cardiac anesthesia. However, over the past few years other monitoring devices have been developed which are now used extensively during open heart procedures. These include: (1) mixed venous oxygen saturation and right ejection fraction monitoring via modified pulmonary artery catheters; and (2) transesophageal echocardiography.

MIXED VENOUS OXYGEN SATURATION

Mixed venous oxygen saturation can be monitored continuously by reflection spectrophotometry with the use of a fiberoptic pulmonary artery catheter. Light of two specific wavelengths, one of which is sensitive to changes in oxygen saturation of hemoglobin and the other which is used as a reference, is transmitted to the pulmonary artery blood, reflected, then transmitted back to a photodetector. The oxygen saturation of this blood, which represents mixing of blood from the coronary sinus, pulmonary artery, and the superior and inferior vena cavae, is then measured as the absorption ratio of the wavelengths.

Continuous mixed venous oxygen saturation monitoring is useful for the evaluation of oxygen consumption relative to supply. It therefore provides useful information regarding tissue oxygenation

At the resting metabolic state the SvO_2 is normally 75%. Under anesthesia the SvO_2 may rise to 85% as oxygen consumption substantially diminishes. Decreases in SvO_2 during surgery can be due to low cardiac output, as tissue supply is reduced in relation to demand, or to insufficient anesthesia, where the oxygen consumption increases relative to supply.

Cardiothoracic Surgery, Third Edition, edited by Fritz J. Baumgartner. ©2004 Landes Bioscience.

RIGHT VENTRICULAR EJECTION FRACTION (RVEF)

RVEF can be measured with the use of a pulmonary artery catheter which contains a rapid response thermistor coupled with an EKG signaling system. Two reference points are detected on the cardiac output thermodilution curve, and the slope of the portion of the curve that connects these reference points is then divided by the R-R interval. The ratio of temperature change per heart beat, which indicates the residual ejection fraction, can then be determined. The right ventricular end diastolic (RVEDV) and end systolic volumes (RVESV) are then calculated as follows:

$$RVEDV = \frac{\text{Stroke volume}}{\text{ejection fraction}}$$

$$RVESV = RVEDV - \text{stroke volume} *$$

$$*\text{stroke volume} = \frac{\text{cardiac output}}{\text{heart rate}}$$

RVEF/RVEDV/RVESV monitoring is useful as intraoperative heart failure is often due to right ventricular dysfunction. In addition, the administration of fluid and cardiac drugs can be tailored according to right ventricular function.

TRANSESOPHAGEAL ECHOCARDIOGRAPHY (TEE)

Transesophageal echocardiography is an extremely useful diagnostic monitor which provides two dimensional imaging of the heart in real time. This is accomplished by one or more transducers placed at the tip of an endoscopic probe which is introduced into the esophagus to the level of the heart. The single or monoplane transducer probe permits transverse imaging. The biplane probe, which contains stacked transducers or an interdigitated type transducer, provides imaging of both transverse and longitudinal planes. The multiplane, or omniplane, instrument allows imaging of multiple planes. Color flow imaging and spectral Doppler flow evaluation, as well as one dimensional time motion mode (M mode) imaging, can also be achieved with TEE.

PROPERTIES OF ULTRASOUND

Ultrasound is composed of pressure waves characterized by their velocity (v), wavelength (l), and frequency (f). The relationship of these three parameters is v=fl, where v=1540m/sec in soft tissues. The frequency of ultrasound waves is greater than 20,000 Hz, making them inaudible to the human ear. Transducer frequencies of 3.5 MHz or 5 MHz are generally used in echocardiography.

As the ultrasound beam wave traverses a medium its intensity will attenuate through absorption, reflection, or scatter. Most of the energy which is absorbed is converted to heat.

Reflection is the redirection of sound in a single direction back to the medium where it originated after striking an interface, such as a blood-tissue interface of a cardiac structure. The physical dimensions of blood-tissue interfaces in the heart are large with smooth, flat surfaces. Imaging is based primarily on sound reflection.

Scatter results when the sound beam is redirected in many directions after striking an interface which has small physical dimensions with a rough surface. The liver and blood stasis in a cardiac chamber may be imaged through scattering.

DOPPLER FLOW

The Doppler principle of flow is expressed by the following relationship:
$V=cDf2f_tCosq$
where V is blood flow velocity;
Df is the frequency shift between reflected and emitted signals;
f_t is the frequency of the emitted ultrasound;
q is the angle between the ultrasound beam and the direction of blood flow;
c is the velocity of sound (1450 m/sec in soft tissues).
The echocardiography machine can display both continuous wave and pulsed wave Doppler representations of blood flow across valves or through vessels.

Continuous wave Doppler (CwD) employs continuous pulses of sound waves, and thus the exact depth of flow cannot be determined. However, blood flows at high velocities can be measured accurately.

Pulsed wave Doppler (PwD) uses short bursts of sound which permits precise location of flow. However, blood flows at high velocities cannot be determined accurately. Instead, at high velocities a phenomenon called aliasing, or wrap-around, occurs with PwD.

For CwD and PwD flow above the baseline indicates flow towards the transducer whereas flow below the baseline indicates flow away from the transducer. However, for PwD, high velocity flow will be displayed on both sides of the baseline (aliasing or wrap-around), regardless of the real direction of flow.

COLOR FLOW DOPPLER

Color flow demonstrates the location, direction, and velocity of cardiac blood flow. Thus, flow acceleration can be estimated and laminar and turbulent flow differentiated.

Two basic colors are assigned according to the direction of flow. Red color indicates flow towards the transducer and blue color indicates flow away from the transducer. When laminar flow accelerates rapidly the intensity of the respective colors increases concomitantly. A mosaic pattern of colors is produced when flow is turbulent.

2-DIMENSIONAL IMAGING

Multiplane TEE is now the most commonly used technology. The multiplane probe can scan the entire field from 0° to 180°, and thus a thorough evaluation of the structures and functions of the heart, including the valves, and the great vessels can be accomplished.

1. The transverse position (at 0°) is used to image: the basal short axis of the great vessels, the aortic valve, and the left atrial appendage.
2. The long axis of the of the heart at various levels to view the chambers, the interventricular septum, tricuspid and mitral valves, the left ventricular outflow tract, and the coronary sinus.
3. The transgastric short axis of the heart is used to view the mitral and tricuspid valves and the left and right ventricles at the basal level, the papillary muscles at the mid short axis level, and the left ventricular outflow tract, ascending aorta, arch, and aortic valve at the deep transgastric level.
   The distal aortic arch and the descending thoracic aorta (short axis). The short axis view of the descending thoracic aorta will also image a left pleural effusion.
4. The multiplane scan (>0°to about 160°) is used to image:
   The short axis at the level of the aortic arch, which also images the pulmonary artery, the pulmonic valve and the left brachiocephalic vein.
   The left side of the heart, including the left atrium, left ventricle, left atrial appendage, left upper pulmonary vein, mitral and aortic valves, left ventricular outflow tract, ascending aorta, and coronary sinus.
   The right side of the heart, including the right atrium, right ventricle, tricuspid and pulmonary valves, right ventricular outflow tract and pulmonary artery.
   The short axis view at the level of the aortic valve, at which the aortic valve, coronary ostia, pulmonic valve and the interatrial septum are imaged.
   The long axis view at the level of the aortic valve, at which the proximal ascending aorta and right pulmonary artery can be imaged in addition to the left ventricular outflow tract and aortic valve.
   The bicaval view, where the right and left atria, the superior and inferior vena cavae and interatrial septum are imaged.
   The long axis of the ascending aorta, which images the right pulmonary artery as well as the ascending aorta.
   The long axis of the descending aorta, which will image left pleural effusions.
   The transgastric two chamber view, imaging the left atrium and ven-

tricle, mitral valve, chordae, and papillary muscles; the transgastric long axis view imaging the left ventricular outflow tract and the aortic and mitral valves; and the transgastric view of the right ventricular inflow, imaging the right atrium and ventricle, the tricuspid valve and chordae, and papillary muscles.

TEE is used routinely during cardiac valve repair/replacement surgeries to assess the degree of valve damage and malfunction immediately before and the presence of residual abnormalities after the procedure. TEE is also used frequently to continuously monitor regional wall motion and diastolic function.

4

ANESTHETIC AGENTS

Various anesthetic techniques can be employed during cardiac surgery. The most common techniques utilize synthetic opiates or inhalational agents, but other agents may be included in the armamentarium. Whatever the technique, one must be aware that all anesthetic agents depress ventricular function to some degree, primarily by their inhibitory effects on calcium ion influx across the sarcolemma.

SYNTHETIC OPIATES

The most common anesthetic agents used during cardiac surgery are the opiates, specifically fentanyl and sufenta. These are synthetic compounds which have relatively little effect on ventricular function and thus maintain hemodynamic stability. However, large doses of fentanyl or sufenta may cause severe bradycardia, particularly if the patient is receiving beta blockers or calcium channel blockers. This bradycardia is most likely due to parasympathetic simulation. Large doses of these opiates can, in the absence of muscle relaxants, cause thoracic muscle rigidity and the resultant "stiff chest syndrome" which may result in difficulties with controlled ventilation. This muscle rigidity is thought to be mediated by the central nervous system.

INHALATIONAL AGENTS

Inhalational agents currently available are sevoflurane, desflurane, isoflurane, halothane and enflurane.

Sevoflurane is the newest inhalational agent. It has a low solubility and thus has a quick onset and disappearance. It is particularly useful for anesthetic induction of children.

Desflurane also has a low solubility, but because it causes irritability of the upper airway it is generally not used for induction. When used at low concentration desflurane has little effect on ventricular function.

Halothane is generally used only for anesthetic induction of pediatric patients. It may cause arrhythmias by sensitization of the conduction pathways of the heart to endogenous or exogenous catecholamines.

Isoflurane may cause significant depression of the right or left ventricular if used at high concentrations for a prolonged period of time. In additional, isoflurane is a vasodilator and may therefore directly reduce systemic vascular resistance. This agent has also been shown to cause coronary steal by diverting blood from ischemic areas of the myocardium toward nonaffected areas, most likely by selective dilatation of the small resistance vessels. When used judiciously with careful monitoring, however, isoflurane should not adversely affect cardiac function.

OTHER ANESTHETICS

Barbiturates, particularly sodium thiopental, are occasionally used as induction agents for open heart surgery. They significantly decrease ventricular function at large doses, by inhibiting calcium ion transport across the sarcolemma. In addition, sodium thiopental causes venodilatation. This venodilatory effect may be useful in situations where reductions in preload are essential, such as in congestive heart failure.

Etomidate, another induction agent, has a much less direct effect on the ventricle than do the barbiturates. It does, however, inhibit steroidogenesis, which generally has little consequence on cardiac function.

Benzodiazepines, including midazolam, do not in themselves have significant effects on cardiac function. However, the concomitant administration of benzodiazepines and high dose fentanyl may result in myocardial depression, particularly if the patient is receiving calcium channel blockers. They cause some vasodilatation.

Propofol, an intravenous sedative-hypnotic anesthetic, is sometimes used as both an induction and maintenance agent for cardiac surgery. Propofol inhibits ionized calcium influx and thus depresses myocardial function. It also produces venous and arterial vasodilatation. Propofol, which has a chemical structure similar to that of the antioxidant vitamin E, has been found to inhibit lipid peroxidation in hepatic and cerebral cells in vitro. Its effects on lipid peroxidation in the myocardium have not yet been studied. Based on these studies, however, propofol, as a result of its antioxidant properties, may be a useful agent for attenuation of postischemic reperfusion injury of various organs in the postcardiopulmonary bypass period of the surgery.

Ketamine, a phencyclidine derivative, is used primarily as an induction agent for children undergoing cardiac surgery. It maintains hemodynamic stability through its sympathomimetic properties and inhibition of the reuptake of norepinephrine. However, by itself, ketamine depresses myocardial function through inhibition of Ca^{2+} flux and may therefore result in severe cardiac decompensation in patients with autonomic failure. Ketamine does have hallucinogenic effects which can be attenuated with the prior use of a benzodiazepine.

Droperidol, a butyrophenone, is a sedative which produces mental detachment. It blocks the transmission of norepinephrine (which makes the agent useful for the control of blood pressure in hypertensive patients), dopamine (which precludes its use in patients afflicted with Parkinson's disease), serotonin and gamma aminobutyric acid in the central nervous system. Large doses of this drug may produce a delayed awakening.

MUSCLE RELAXANTS

Muscle paralysis is a necessary component of cardiac anesthesia. There are currently nine relaxants which are nondepolarizers (i.e., they block acethycholine release at the neuromuscular junction) and one which is a depolarizer (i.e., depolarizes the neuromuscular junction).

Several nondepolarizers are used for cardiac anesthesia. Many have systemic effects which could be either detrimental or helpful for patients undergoing open heart surgery. The older nondepolarizers include tubocurarine, metocurine, and pancuronium. These relaxants are long acting. Tubocurarine and metocurine block autonomic ganglia and cause histamine release. Pancuronium blocks parasympathetic receptors and has some sympathogonic activity, actions which may be useful in patients who present with significant bradycardia due to beta or calcium channel blockers.

Newer nondepolarizers have been developed which have little or no hemodynamic effects. Atracurium and vecuronium have an intermediate duration of action.

Atracurium is degraded through the disruption of this chemical structure (Hofman elimination pathway) and ester hydrolysis in plasma, which makes this relaxant useful in patients with kidney or liver failure. However, atracurium does induce histamine release. Cisatracurium is an isomer of atracurium. It appears to cause less histamine release than does atracurium.

Vecuronium has no hemodynamic effects, but it is metabolized by the liver and is excreted by the kidney and bile.

Mivacurium is a short acting nondepolarizer with virtually no hemodynamic effects. It is useful for induction of anesthesia.

Rocuronium is a long acting nondepolarizer which, at large doses, has a rapid onset of action. At a dose of 1.2 mg/kg Rocuronium can provide paralysis sufficient for intubation in less than 90 seconds.

The depolarizer, succinylcholine, is very short acting and is generally used only in emergency situations where the airway must be immediately secured by intubation or in cases where preintubation ventilation during induction of anesthesia is very difficult because of an abnormal airway or chest wall rigidity due to large doses of fentanyl or sufenta. Muscle fasciculations occur after injection which reflect the depolarizing activity of succinylcholine. Postoperative muscle pain may result from fasciculation. Administration of a small dose of a nondepolarizer prior to the administration of succinylcholine may prevent these fasciculations.

Table 4.1. Cardiac Drugs—Antiarrhythmic Drug Therapy

Class	Mechanism of Action	Common Drugs Used in the OR
I A	Blocks sodium channels	Procainamide
	Blocks different potassium channels	
I B	Blocks sodium channels	Lidocaine
II	Indirectly blocks opening of the calcium channel by blunting sympathetic activation	Beta blockers (esmolol)
III	Blocks potassium channels	Amiodarone, bretyllium, ibutilide
IV	Directly blocks calcium channels	Verapamil, diltiazem, nicardipine

Amiodarone is now the first line drug for the treatment of recurrent and destabilizing ventricular tachycardia. It lowers the threshold for defibrillation (lidocaine increases this threshold). Like any drug that blocks any part of the potassium channel (and thus prolongs the refractory period), amiodarone can cause atypical ventricular tachycardia (torsade de pointe).

Adenosine is the treatment of choice for supraventricular tachycardia. It binds to purinergic receptors which interact with the potassium channels, slowing conduction time through the AV node. It also interrupts reentry pathways through the AV node. Adenosine is taken up by erythrocytes and vascular endothelial cells. Its activity is potentiated by nucleoside transport inhibitors such as dipyrdamole and inhibited by adenosine receptor antagonists such as the methylxanthines.

Cardioversion is the treatment of choice for atrial fibrillation. Ibutilide is an alternative pharmacological treatment for atrial fibrillation; however, torsade de pointe is a frequent complication of ibutilide.
Automatic Implantable Cardioverter Defibrillators (AICD) are now commonly implanted in patients with unsustained, spontaneous ventricular tachycardia with left ejection fractions of between 20 and 40%.

Table 4.2. Cardiac Drugs—Antiarrhythmic Drug Dosages

Adenosine	6 mg IVP, 12 mg IVP in 1-2 min.: Peds 50 µg/kg IVP
Amiodarone	Loading dose: 150 mg over 10 minutes; then 360 mg over the next 6 hours; then 540 mg over the next 18 hours
Bretylium	Loading 5-10 mg over 8-10 min, maint. 5-10 mg q6h
Esmolol	.2-.5 mg/kg bolus; maint. .05-.1 mg/kg/min
Ibutilide	.5-1 mg infusion over 10 min (.005-.01 mg/kg if patient weighs less than 60 kg
Lidocaine	Loading 1-1.5 mg/kg, maint. .1 mg/kg/hr
Procainamide	1 gram over 30 minutes; maint. 1-6 mg/min: Peds 3-6 mg/kg over 5 minutes (max 100 mg/dose)
Verapamil	5-10 mg over 2-3 minutes, may repeat dose 15-30 min. Peds. .1-.2 mg/kg over 2 minutes

Succinylcholine stimulates parasympathetic receptors and autonomic ganglia and can therefore cause arrhythmias. It may also induce histamine release. Other complications due to succinylcholine include hyperkalemia and malignant hyperthermia in susceptible patients.

Table 4.3. Cardiac Drugs—Inotropic Drugs

Sympathomimetic Drugs

Dobutamine (dobutrex)	250 mg/250 ml (1000 µg/ml) 2-20 µg/kg/min
Dopamine (intropin)	400 mg/250 ml (1600 µg/ml) 3-10 µg/kg/min
Ephedrine	5-25 mg IVP
Epinephrine	2 mg/250 ml (8 µg/ml) 1-8 µg/min, 1 µg/min = 7.5 ml/hr
Isoproterenol (isuprel)	1 mg/250 ml (4 µg/ml) .05-.1 µg/kg/min, 1 µg/min = 15 ml/hr
Norepinephrine (levophed)	8 mg/100 ml (80 µ/ml) .05-.3 µ/kg/min

Phosphodiesterase Type III Inhibitors

Amnirone (inocor)	100 mg/250 ml (400 µg/ml) 1-1.5/kg loading, 5-20 µ/kg/min
Milrinone (primacor)	100 mg/500 ml D5 (200 µg/ml) 50 m/kg loading over 10 min., .375-.75 µg/kg/min maintenance

The phosphodiesterase inhibitors should be used in chronic congestive heart failure or other conditions when the β receptors are down regulated.

Other Inotropes

Calcium chloride	10-20 mg/kg, 250-1000 mg IV
Calcium gluconate	30-60 mg/kg IV
Digoxin (lanoxin)	1-1.25 mg (.25-.5 mg increments) or 10-15 µg/kg loading dose, .125-.25 mg/day

ANTIARRHYTHMIC, INOTROPIC AND VASOPRESSOR THERAPY

Antiarrhythmic drug therapies are classified in Table 4.1, and 4.2 lists their dosages. Inotropic and vasopressor treatments are summarized in Tables 4.3 and 4.4 respectively.

COAGULATION

BASIC MECHANISMS OF COAGULATION

The coagulation process is activated by both an intrinsic and extrinsic system.

The intrinsic, or contact, system is a cascade of reactions which require the formation of complexes between surface bound enzyme and cofactors. The intrinsic cascade begins with the complexing of three plasma proteins, high molecular weight kininogen, pre-kallikrein, and factor XII, with subendothelial collagen. Factor XII is then converted from an inactive to an active protease (XIIa). The cascade then proceeds as follows: activated factor XIIa->factor XIa->factor IXa.

The extrinsic system is mediated by a cell membrane lipoprotein, tissue factor, which complexes to factor VII and calcium. Thus, factor VII is converted to an active protease.

The reactions of the intrinsic and extrinsic pathways generate proteases which activate factor X. At this level prothrombin is converted to thrombin. Thrombin then converts fibrinogen to fibrin. Plasmin, which is activated from plasminogen

Table 4.4. Cardiac Drugs—Vasopressors

Phenylephrine (Neo-synephrine)	100-300 m IVP; 30 mg/250 ml (120 m/ml), 10-500 m/min, 30 m/min = 15 ml/hr
L Vasopressin (Pitressin)	100 units/100 ml (1 units/ml) .04-.1 units/min

L Vasopressin is useful in cases of vasodilatory shock, such as that seen post cardiopulmonary bypass.

Vasodilators

Nitroglycerin	50 mg/250 ml (200 mg/ml), 1-4 m/kg/min
Nitroprusside	50 mg/250 ml (200 mg/ml), 1-2 m/kg/min
Phentolamine	10 mg/100 ml (100 mg/ml) fast infusion
Prostaglandin E1	.05 mg/kg/min
(PGE1, Prostin, Alprostadil)	

Prostaglandin E1 is indicated primarily for pulmonary hypertension, and it is thus infused through a right atrial line. Concomitant infusion of a vasopressor through a left atrial line is usually necessary to maintain systemic pressures.

by tissue plasminogen activator, degrades fibrin to low molecular weight fragments.

Antithrombin, protein C and protein S regulate the coagulation system.

Platelet function is regulated by the von Willebrand's factor, which binds the glycoprotein Ib receptor of the platelet to the vessel wall, and by fibrinogen, which binds platelets to each other through their respective glycoprotein IIb-IIIa receptors.

COAGULATION PROFILE

Aside from such tests as the partial thromboplastin time (PTT), prothrombin time (PT), and thrombin time (TT), which examine the intrinsic, extrinsic, and fibrinogen to fibrin pathways, respectively, the coagulation profile can be assessed with the use of the thromboelastograph (TEG) or viscoelastograph (Sonoclot).Both the TEG and the Sonoclot record the entire coagulation process.

The TEG measures coagulation with a cuvette and piston which become coupled as a clot is formed. The shearing elasticity of this clot is translated onto thermal paper from the cuvette-piston coupling. The TEG thus records a:

1. reaction time, which corresponds to the rate of generation of thromboplastin;
2. coagulation time, which measures the time for a clot to begin to form;
3. clot formation rate, which measures the time for a solid clot to form;
4. maximum amplitude, which is related to clot elasticity; and
5. whole blood lysis time, which corresponds to fibrinolysis.

The Sonoclot represents the coagulation process as a "signature". A piston is used which detects the viscous drag as a clot is forming. This drag is quantified electronically as a change in electrical output. The resultant signal is then translated onto a chart as a tracing, or signature. The Sonoclot thus records:

1. time of onset of fibrin formation, mediated by the both the intrinsic and extrinsic pathways. This is the activated clotting time of the Sonoclot (SonACT).

2. the rate of conversion from fibrinogen to fibrin;
3. fibrin-platelet interaction; and
4. the rate of clot contraction, which is the rate at which the platelets contract the fibrin clot.

PHARMACOLOGIC INTERVENTIONS

Three categories of procoagulant drugs are currently used to reduce perioperative bleeding: desmopressin (DDAVP), lysine analogues (epsilon aminocaproic acid and tranexamic acid), and serine protease inhibitor (aprotinin).

Desmopressin is a synthetic analogue of the antidiuretic hormone. It increases the activity of factor VIII and interacts with the glycoprotein 1b receptor on the platelet to induce adhesion of the platelet to the vascular subendothelium. It is effective in the treatment of bleeding diathesis in patients with renal failure, cirrhosis, or von Willebrand disease.

The lysine analogues exhibit antifibrinolytic activity by binding to a lysine site on plasminogen. In this way they inhibit activation of the plasminogen and thereby its interaction with fibrin.

The serine protease inhibitor aprotinin has been shown to be extremely effective in reducing bleeding. It blocks various mediators such as kallikrein and plasminogen and thus acts as both an antiinflammatory and an antifibrinolytic agent. Complement activation is also inhibited. Aprotinin protects platelet integrity by preventing destruction of the glycoprotein IIb-IIIa receptor by plasmin. These properties make aprotinin a very effective agent for open heart surgeries as it substantially reduces bleeding and thus the need for blood and blood product transfusions, particularly platelets which may in themselves aggravate complement activation. Because of its antiinflammatory effects aprotinin may also reduce the need for inotropic support in the post cardiopulmonary bypass period.

Severe anaphylactic reactions to aprotinin may occur if it had been administered previously within six months. Aprotinin rarely causes renal dysfunction or thrombosis even when used during circulatory arrest.

THORACIC ANESTHESIA

Surgical interventions which involve thoracotomy and lateral positioning of the patient generally require that the ipsilateral lung remain deflated to optimize exposure. One lung ventilation is accomplished by the use of a double lumen tube (DLT) which consists of an endobronchial tube as well as a tube which remains in the trachea. Each tube is cuffed. Right and left DLTs are designed so that the endobronchial tube enters the right or left bronchus, respectively.

It is essential that the endobronchial tube be placed correctly in the bronchus. Confirmation of correct placement is accomplished by auscultation, direct visualization and observation of ventilatory parameters.

AUSCULTATION

One of the tubes is clamped. Breath sounds should then be heard upon auscultation only on the side corresponding to the patent tube. Moreover, only the corresponding hemithorax should rise upon positive pressure ventilation.

DIRECT VISUALIZATION

Fiberoptic bronchoscopy is used to visualize placement of the DLT. The bronschoscope is introduced through the tracheal tube and is advanced to its outlet. At this level the carina should be clearly visible. In addition, the cuff of the endobronchial tube should be visible in the bronchus. Chest radiography is not as precise as fiberoptic bronchoscopy for confirmation of tube placement.

VENTILATORY PARAMETERS

Adequate one lung ventilation can be confirmed after placement of the DLT. The tube which is ventilating the lung that will ultimately be deflated during the surgical procedure is clamped. After establishment of one lung ventilation the end-tidal carbon dioxide, peak inspiratory pressure and oxygen saturation should not change significantly. Any significant decrease in end-tidal CO_2 and increase in peak inspiratory pressure indicate airway obstruction and require repositioning of the DLT. A decrease in O_2 saturation may indicate airway obstruction or severe atelectasis. An arterial blood gas should be analyzed 10 minutes after one lung ventilation has been initiated.

The DLT can be dislodged from its correct position during positioning of the patient or during surgical manipulation, leading to airway obstruction and desaturation. Repositioning of the DLT with the aid of a fiberoptic bronchoscope should be attempted. Efforts to improve oxygenation include providing continuous positive airway pressure to the deflated lung or, if the end-tidal CO_2 remains within normal limits, applying positive end expiratory pressure to the lung being ventilated. The DLT should be periodically suctioned to avoid obstruction by secretions or blood.

Left DLTs are used for procedures necessitating a right thoracotomy. Many anesthesiologists, concerned that a right DLT may not provide adequate ventilation to the right upper lobe (because of the possibility of occlusion of the takeoff of the right superior lobe bronchus by the endobronchial tube), advocate the use of a left DLT for procedures requiring a left thoracotomy as well. However, DLTs designed for right endobronchial intubation contain a large side opening which easily opposes this takeoff. Patency of the right superior lobe bronchus, therefore, is generally maintained with right DLTs. Ventilation of the right upper lobe can be confirmed by fiberoptic bronchoscopy and by auscultation of the axilla.

Intubation of the left bronchus for procedures involving the left hemithorax (and thus requiring deflation of the left lung) may also lead to complications in certain situations. A bronchus rendered stiff by an endobronchial tube can be injured during surgical manipulation. Moreover, ventilation to the right lung may be compromised as mediastinal contents may compress the unintubated right bronchus when the patient is placed in the right lateral decubitus position. Left DLTs must be inserted with caution in the presence of a thoracic aneurysm; abrupt intubation of the left bronchus can rupture an overriding aneurysm.

An alternative to the double lumen tube is the wire-guided endobronchial blocker. The wire-guided endobronchial blocker is particularly useful in a patient with a difficult airway, which would make insertion of a double lumen tube extremely difficult, or in a patient requiring nasal intubation, which would preclude the use of a double lumen tube.

4

SUGGESTED READING

1. Benumof J, Alfery D. Anesthesia for thoracic surgery. In: Miller, ed. Anesthesia. New York: Churchill Livingstone, 1994:1689-1700.
2. Birman H, Haq A et al. Continuous monitoring of mixed venous oxygen saturation in hemodynamically unstable patients. Chest 1985; 85(5):753-756.
3. Dhainaut JF, Brunet F et al. Bedside evaluation of right ventricular performance using a rapid thermodilution method. Crit Care Med 1987; 15(2):148-152.
4. Hunger J. Drug therapy: New neuromuscular blocking drugs. N Engl J Med 1995; 332(25):1691-1698.
5. Landry DW, Oliver JA. Mechanisms of disease: The pathogenesis of vasodilatory shock. N Engl J Med 2001; 345:588-595.
6. Musacchio E, Rizzol V, Bianch M et al. Antioxidant action of propofol on liver microsomes and brain synaptosomes in the rat. Pharmacology 1991; 69(1):75-77.
7. Pagel P, Grossman W, Haering T et al. Left ventricular diastolic function in the normal and diseased heart. Anesthesiology 1993; 79(5):1104-1120.
8. Tardif JC et al. Transesophageal echocardiography: Current instrumentation and future prospects. In: Maurier, ed. Transesophageal Echocardiography. McGraw-Hill, Inc., 1994: 5-8.
9. American Heart Association. Guidelines 2000 for cardiopulmonary resuscitation and cardiovascular care. Circulation 2000; 102(suppl):I-86-I-166
10. Bigger JT Jr. Antiarrhythmic drugs in ischemic heart disease. Hosp Pract 1972; 7:69.
11. Dorian P, Cass D, Schwartz B et al. Amiodarone as compared with lidocaine for shock-resistant ventricualr fibrillation. N Engl J Med 2002; 346:884-890.
12. Fozzard HA, Gibbons WR. Action potential and Contraction of heart muscle. Am J Cardiol 1973; 31:182.
13. Katz AM. Cardiac ion channels. N Engl J Med 1993; 328(17):1244-1251.
14. Katz AM. Physiology of the heart. 2nd ed. New York: Raven Press, 1992:415-541.
15. Moss AJ, Jackson Hall W, Cannom DS et al. Improved survival with an implanted defibrillator in patients with coronary disease at high risk for ventricular arrhythmia. N Engl J Med 1996; 335(26):1933.
16. Moss AJ, Zareba W, Hall WJ et al. Prophylactic implantation of a defibrillator in patients with myocardial infarction and reduced ejection fraction. N Engl J Med 2002; 346:877-883.
17. Rosen MR. Electrophysiology and pharmacology of cardiac arrhythmias. Cardiac antiarrythmic effects of lidocaine. Am Heart J 1975; 89:526.
18. Sokolow M, McIlroy M.B. Conduction Defects. In: Clinical Cardiology. 3rd ed. Pathway in Wolff-Parkinson White syndrome. Br Heart J 1975; 37:127.

19. Vaughan Williams EM. Classification of antiarrhythmic drugs. In: Sandoe E, Flensted-Jensen E, Olesen KH, eds. Symposium on cardiac arrhythmias. Elsinore, Denmark: Astra, 1970:449-472.

20. Goldman M. Examination techniques, anatomy, and basic views. In: Goldman M, ed. Clinical Atlas of Transesophageal Ecocardiography. 1st ed. Futura Publishing Co., Inc., 1993:5-8.

21. Hedrick W, Hykes D, Starchman D. Ultrasound Physics and Instrumentation. 3rd ed. Mosby-Year Book, Inc. 1995.

22. Oh JK, Seward JB, Tajik AJ. The Echo Manual. 2nd ed. Philadelphia: Lippencott-Raven 1999:112, 122.

23. Rafferty T. TEE imaging: Basic transverse plane examination sequence. Video J Echocardiogr 1997; 7(3):39-56.

24. Seward JB, Khandheria BK, Freeman WK et al. Multiplane transesophageal echocardiography: Image orientation, examination technique, anatomic correlations and clinical applications. Mayo Clin Proc 1993; 68:1-29.

25. Shanewise J, Cheung A, Aronson S et al. Guidelines for performing a comprehensive intraoperative multiplane transesophageal echocardiography examination. J Am Soc Echocardiogr 1999; 12(10):884-898.

26. Amicar (aminocaproic acid) injection. Pearl River, NY: Lederle Laboratories Division, 1999 (Physicians' Desk Reference). DDAVP (desmopressin acetate) injection. Collegeville, PA: Rhone-Poulenc Rorer Pharmaceuticals, Inc., 2000 (Physicians' Desk Reference).

27. Furie B, Furie BC. Molecular and cellular biology of blood Coagulation. N Engl J Med 1992; 326(12):800-806.

28. Handin R. Bleeding and thrombosis. In: Braunwald E, Isselbacher KJ, Petersdorf RG et al, eds. Harrison's Principles of Internal Medicine. 11th ed. New York: McGraw-Hill, 1987:266-72.

29. Kang YG, Martin DJ, Marquez J et al. Intraoperative changes in blood coagulation and thromboelastographic monitoring in liver transplantation. Anesth Analg 1985; 64:888-896.

30. LaForce WR, Brudno DS, Kanto WP et al. Evaluation of the Sonoclot Analyzer for the measurement of platelet function in whole blood. Ann Clin Lab Sci 1992; 22:30-3.

31. Royston D, Nadel A, Dietrich MD et al. Aprotinin use and adverse outcomes associated with platelet administration. Anesth Analg 2000; 90:SCA19. Sienco, Inc. Sonoclot; Coagulation User Manual 1991.

32. Tuman KJ, Spiess BD, McCarthy RJ et al. Comparison of viscoelastic measures of coagulation after cardiopulmonary bypass. Anesth Analg 1989; 69:69-75. Trasylol (aprotinin) injection. West Haven, CT: Bayer Corporation, 2000 (package insert).

33. Van Oeveren W, Harder MP, Roozendaal KJ et al. Aprotinin protects platelets against the initial effect of cardiopulmonary bypass. J Thorac Cardiovasc Surg 1990; 99:788-97.

34. Woodman RC, Harker LA. Bleeding complications associated with cardiopulmonary bypass. Blood 1990; 76:1680-97.

Coronary Artery Disease

Fritz J. Baumgartner and Matthew Budoff

EVALUATION OF THE CORONARY PATIENT

The patient who presents with chest pain needs to be thoroughly evaluated. A complete history and physcial is first undertaken, taking special care to note the presence of cardiac risk factors: history of premature family history of CAD, hypertension, diabetes, obesity, hypercholesterolemia and smoking. A physical examination should focus on the cardiovascular system, followed by laboratory assessment of cardiac enzymes (troponin, CPK and MB), a chest x-ray and electrocardiogram. Suspicion of an acute coronary syndrome (unstable angina or acute myocardial infarction) requires oxygen therapy, nitrates and aspirin therapy (unless contraindicated), blood pressure therapy (first line therapy of acute coronary syndromes is beta blockers) and serial electrocardiograms and cardiac enzymes. Evidence of an acute myocardial infarction by either enzyme analysis or electrocardiogram usually requires immediate therapy with either thrombolytics or angioplasty. Currently there are several thrombolytics on the market, and utilization varies depending upon multiple factors including the location of infarction. Furthermore, studies demonstrate reduced morbidity and mortality with other antiplatelet agents (clopidrogrel and IIb-IIIa inhibitors) as well as heparin, either unfractionated or fractionated. The patient, depending upon their status, may be admitted to a chest pain unit (myocardial infarction unlikely) or a coronary care unit (present or likely infarction).

The patient is observed on bedrest and medical therapy. For continued pain or evidence of ischemia, urgent revascularization , either with angioplasty or bypass, is usually indicated. If myocardial infarction can be ruled out on the basis of normal serial enzymes and no evolution of EKG, the patient most often is tested with cardiac stress testing (treadmill, stress echocardiogram or nuclear testing). See Chapter 2 for cardiac testing. Patients who pass this noninvasive evaluation (no ischemia or chest pain) are most often discharged with a diagnosis of noncardiac chest pain.

If the patient has a positive stress test (ST depression or chest pain), the patient is most often referred for cardiac catheterization (see Chapter 2).

UNTREATED SURVIVAL

SURVIVAL WITHOUT SURGICAL TREATMENT ACCORDING TO ANATOMY

For one vessel disease, the survival after 5 years is the same as normal (except for isolated proximal LAD lesions).

For two vessel disease, there is a 75% 5 year survival compared to 95% of the general population. For three vessel disease, there is a 50% 5 year survival compared to 95% survival for the age matched general population. For left main coronary artery disease, the survival without treatment is similar to triple vessel disease.

SURVIVAL WITHOUT TREATMENT ACCORDING TO LEFT VENTRICULAR FUNCTION IN CORONARY ARTERY DISEASE

In patients with coronary disease, for normal ejection fractions, the 5 year survival is 92%; for ejection fractions between 30-50% the 5 year survival is reduced to 75%.

As a general rule, the number and severity of atherosclerotic lesions progress with time. Patients with unstable angina have a decreased life expectancy compared to stable angina. One can therefore conclude that coronary revascularization will best help those with reduced ejection fractions and those with unstable angina. Although the presence of reduced ejection fraction in coronary patients is the best prediction of those whose cardiac function will most benefit by surgery, this factor also has the greatest chance of resulting in operative mortality. In terms of relief of angina after coronary artery bypass grafting, 80-90% of patients remain relieved of their angina at 5 years; 50% are relieved of their angina at 10 years. After medical treatment, 3% of patients are relieved of their angina at 10 years showing the superiority of coronary revascularization compared to medical treatment for relief of angina.

UNSTABLE ANGINA

Unstable angina is a condition of chest pain which is rapidly progressive or alternatively is pain at rest and frequently progresses to myocardial infarction. The Prinzmetal's variant angina is coronary artery spasm usually superimposed on atherosclerotic vessels but may occur in normal vessels. Unlike stable or unstable angina, Prinzmetal's angina may not necessarily be brought on by anxiety or effort. ST segments in Prinzmetal's angina are elevated. The pain is often relieved by myocardial infarction while pain in stable angina is often increased after myocardial infarction.

Soon after the onset of symptoms in Prinzmetal's angina, the natural history is often a catastrophic event, i.e., myocardial infarction or death soon after the onset of symptoms. Patients with Prinzmetal's angina are treated with nitrates. Ergot alkaloids for headaches are avoided as is exposure to cold. Calcium channel blockers are very helpful in addition to nitrates. Coronary revascularization may be useful for medically refractory patients who have atherosclerotic disease with superimposed Prinzmetal's angina but is not useful if there is no significant atherosclerosis present.

Patients who present in cardiogenic shock after an acute myocardial infarction require intensive management. Their airway is assessed and if necessary, they are intubated. IVs are started with blood drawn for enzyme analysis and electrolytes. Chest x-ray and EKG are done and a Swan-Ganz catheter and Foley catheter and arterial line are placed. The patient is managed pharmacologically with inotropes, afterload reducers and diuretics, as well as nitrates and calcium channel blockers. Beta blockers are used with caution because of cardiogenic shock.

If cardiogenic shock persists, the next line of therapy is an intra-aortic balloon pump. This is used for two reasons: (1) It decreases the afterload against which the heart must work by deflating just prior to systole. Thus, there is decreased myocardial oxygen utilization. The decrease in afterload results in improved ejection fraction and decrease in left ventricular end diastolic pressure and volume, and decrease in workload on the heart. (2) The balloon then inflates in diastole. The coronary artery perfusion pressure is increased, improving oxygen delivery to the myocardium. Thus at precisely the time that the heart needs it the most, there is decreased workload of the heart and increased myocardial oxygen delivery. There is no question that the balloon pump has saved the lives of many patients in cardiogenic shock. Patients in cardiogenic shock should have coronary and cardiac angiography. If a patient will not tolerate a dye ventriculogram because of possible volume overload, a transthoracic echocardiogram may be warranted. A decision is then made whether these patients should undergo coronary artery bypass surgery.

PERCUTANEOUS TRANSCORONARY ANGIOPLASTY (PTCA)

Angioplasty is the cardiologist's method of revascularization; unfortunately it is a poor second to coronary artery bypass surgery and this is now only coming to full light. It is suspected that after angioplasty, up to 50% of lesions are restenosed after 6 months. Angioplasty is said to be initially successful in up to 90% of cases. Angioplasty done on multivessel disease increases the mortality from 1-3%. The overall Q-wave infarction rate for angioplasty is 5%. Contraindications to angioplasty include proximal LAD lesions (proximal to the first septal perforator), left main disease, long segments of disease which would be difficult to get a balloon catheter across, chronic occlusions, circumferential lesions, heavily calcified lesions, bifurcations or the appearance of liquid cholesterol which may embolize. PTCA indications include significant angina or positive exercise test in the presence of 75% narrowing with single or double vessel disease refractory to medical treatment without any of the above contraindications. Acute myocardial infarction with or without prior thrombolytic therapy may be another indication.

It should be noted that a failed angioplasty resulting in the urgent need for coronary artery surgery is a major crisis. Emergency coronary artery bypass surgery after failed PTCA significantly increases the risk of operative death and perioperative MI compared to elective CABG. The risk of operative death and perioperative MI after a de novo coronary artery bypass grafting (CABG) without prior PTCA is 1% and 6%; the mortality of a coronary artery bypass surgery after a failed PTCA increases to 5-10%, and the risk of perioperative myocardial infarction (Q-wave myocardial infarction) increases to 25%. After a PTCA, the chance of a failure requiring an emergency coronary artery bypass surgery is 5%.

There are several technical details regarding coronary artery bypass surgery after failed PTCA. These include the problems of dissection of the coronary artery due to the PTCA, problems of free rupture of the coronary artery and problems of distal embolization. The dissection may be managed by being sure that the open coronary lumen is in fact a true lumen. This can be determined by examining closely the back surface of the lumen. If it is concave, then it is most likely the true lumen; however, if it is convex, then it is most likely the false lumen and the convexed surface will need to be further incised to enter the true lumen. Pre-rupture of the coronary artery is managed by ligation of the artery at the site of perforation and distal coronary bypass grafting. Distal embolization is managed by angiographically being sure where the distal embolization occurs and then performing the coronary bypass distal to this.

In terms of management of the patient in the angiography suite after a failed coronary angioplasty, the following points should be borne in mind:

1. The main cause of morbidity and mortality from coronary bypass surgery after failed PTCA is a delay in coronary bypass surgery after the PTCA. The main cause of delay is persistence on the part of the cardiologist to maintain an open vessel past the area of iatrogenic injury by further instrumentation. It is an important point that there should be no further delay. The only instrumentation that should be performed in the cath lab after a PTCA accident is to place a "bail-out" catheter which contains multiple holes proximal and distal in the catheter to perfuse past the site of obstruction.

2. Persistent ST segment elevation, persistent chest pain refractory to nitrates, calcium channel blockers and beta blockers, or hypotension after a failed PTCA warrants placement of an intra-aortic balloon pump (in the catheterization lab in preparation for surgery).

3. The most important factor to improve survival in patients undergoing a failed PTCA is urgent coronary bypass surgery.

Continuing ongoing unresolved debate exists between medical management versus PTCA versus coronary artery bypass surgery in patients with coronary artery disease. As mentioned, 50% of patients who undergo PTCA can be expected to develop recurrence of a hemodynamically significant lesion requiring reangioplasty or surgery.

Studies in the past have revealed the following caveats regarding these three modalities of management also:

1. *PTCA versus medical treatment in patients with stable and unstable angina and 1-2 vessel coronary artery disease (excluding proximal LAD lesions).*

PTCA significantly increases angina-free survival. No overall difference in survival exists between PTCA versus medical treatment, but survival is significantly better in two subsets undergoing PTCA: those with impaired left ventricular function and those with two-vessel coronary artery disease.

2. *PTCA versus coronary artery bypass in patients with severe proximal LAD stenosis.* Although no difference in survival between these two treatment modalities has been determined, PTCA of an important proximal LAD stenosis may be quite dangerous and may be better treated with CABG.

CORONARY BYPASS SURGERY

5

INDICATIONS

Coronary artery bypass surgery has, over the years, proven its efficacy in terms of improved patient survival and improved quality of life in terms of pain-free survival. The general indications for coronary revascularization include improvement in survival, both overall survival as well as increase in angina free survival. The specific indications for coronary artery bypass grafting are as follows:

1. Triple vessel disease with or without decrease in ejection fraction (but particularly with a decrease in ejection fraction). Patients with reduced ejection fraction with coronary artery disease have a 5 year survival of approximately 75%, this increases to 90% in patients undergoing coronary artery bypass grafting.

2. Double vessel coronary artery disease with reduced ejection fraction. Coronary artery bypass surgery appears to increase long-term survival compared to medical management. However, double vessel disease with normal ventricular function probably can be managed as effectively with angioplasty or medical management.

3. Angina refractory to triple vessel therapy including nitrates, beta blockers and calcium channel blockers.

4. Compelling anatomy; this includes left main coronary artery disease in which the patients have a propensity to sudden death; also proximal left anterior descending artery disease where there is calcification proximal to the first septal perforator. This instance is associated with a high incidence of sudden death and should be managed with angioplasty cautiously if at all.

Life threatening ventricular arrhythmias after myocardial infarction even without a left ventricular aneurysm is an indication for coronary artery bypass surgery. Some type of anti-arrhythmic surgery, either endocardial ablation or placement of an automatic implantable cardioverter defibrillator, should be done as well. Unstable angina, i.e., crescendo angina, is an indication for coronary revascularization on a semi-urgent basis if full medical therapy is ineffective for up to several days. If an EKG shows infarction, urgent coronary artery bypass grafting may be performed or preoperative thrombolytic therapy may be tried.

CONDUITS USED FOR CORONARY REVASCULARIZATION

The two most commonly used conduits for a coronary revascularization are the greater saphenous vein and the internal mammary artery. Over the years, there has been no question that increased use of the internal mammary artery results in improved survival and improvement of angina free survival. This is because of the superior patency of the internal mammary artery graft compared to the saphenous vein graft. The 10 year patency of a saphenous vein graft is on the order of 50%. The 10 year patency of an internal mammary artery graft is on the order of 95%. The saphenous vein graft loses patency for the following reasons: Initially, intimal hyperplasia occurs as a remodeling process for the vein to adapt itself to the artery. This can result in occlusion of the graft. Later on atherosclerosis occurs within the vein graft, and this ultimately leads to the majority of saphenous vein graft occlusions. The internal mammary artery graft has the advantage that it will not occlude by either of these processes since it does not undergo intimal hyperplasia and it does not undergo atherosclerosis. The left internal mammary artery is used most often and may be placed typically onto the left anterior descending artery or onto the obtuse marginal arteries coming off the circumflex. The right internal mammary artery may be used for the anterior descending artery or may be brought to the right coronary artery. It may also be brought down through the transverse sinus to be placed onto the circumflex artery. The mammary artery may also be used as a free conduit by transecting it proximally and attaching it separately onto the aortic root. In general, caution regarding the use of the internal mammary artery should be used in the following situations:

1. Diabetic patients in general may have one internal mammary artery taken down; however, two internal mammary arteries would probably be dangerous because of the decrease in vascularity of the chest wall, resulting in suboptimal wound healing.

2. Immunocompromised patients and patients in chronic renal failure likewise should undergo one internal mammary artery takedown at the most.

3. Extremely old patients who would most likely expire within the next 10 years would likely not benefit from an internal mammary bypass graft.

4. Patients who have atherosclerotic subclavian arteries would most likely not benefit from an internal mammary artery because of the atherosclerosis which may progress to the internal mammary artery, as well as the subclavian artery atherosclerosis which may impair flow to the mammary artery.

5. Patients requiring emergency surgery for cardiogenic shock generally should not be subjected to the increased time it takes for taking down the internal mammary artery.

6. Patients who have a severely calcified or extremely tiny target coronary artery generally would minimally benefit from an internal mammary artery graft because of severity of disease in the distal target.

7. Redo-coronary surgery patients who have had previous vein grafts should receive mammary arteries with caution. In particular, the situation in which a stenotic vein graft to a completely occluded large important coronary artery requires reoperation. In this case, it may not be wise to use the internal mammary artery since the entire blood flow to that large coronary artery depends on the vein graft, and the mammary artery may not be able to supply the large demand of the large coronary artery in the acute postoperative phase. It should be noted that one of the major problems with internal mammary artery grafts is the propensity for these grafts to go into spasm in the perioperative period which can result in acute infarction and hemodynamic destabilization. This is why nifedipine is given intraoperatively after an internal mammary artery is performed, as well as postoperatively.

Other possible conduits for coronary revascularization include the lesser saphenous vein and gastroepiploic artery, as well as inferior epigastric artery. The gastroepiploic and inferior epigastric arteries seem to have comparable patency to the internal mammary artery. The lesser saphenous has a patency comparable to the greater saphenous vein. Arm veins have been used, however, have an extremely poor patency in the order of 50% after 2 years. Radial artery conduits are used as well.

The natural history of vein grafts is that at 2-1/2 years, about a third show significantly decreased flow (i.e., greater than 50% reduction in diameter); about a third show mild reduction; and a third are normal. By 10 years, about half of vein grafts are occluded. Intimal hyperplasia is a remodeling process starting in the vein grafts older than 1 month and results in decreased flow to approximate the caliber of vein with the caliber of recipient artery. The absence of cineangiographically evident intimal hyperplasia at 1 year is a good sign because only 10% of these will later develop intimal hyperplasia on cineangiogram 5 years postoperatively. Part of the pathophysiology of graft closure is due to atherosclerosis causing 1-3% of grafts closing per year. The reason the internal mammary artery is superior to the saphenous vein graft is because no intimal hyperplasia occurs and little to no atherosclerosis occurs, hence the 10 year patency of the internal mammary artery is 95% compared to 50% for the saphenous vein graft.

To decrease the incidence of postoperative graft thrombosis, aspirin and persantine are useful. In one landmark study evaluating the efficacy of aspirin and persantine in decreasing graft occlusion, it was shown that 1 month after coronary artery bypass, patients who had aspirin and persantine had a reduction in their graft occlusion rate from 21-8%. Similarly, at 6 months, the graft occlusion rate was decreased from 38-10%. Aspirin and persantine are, therefore, very effective medications in decreasing graft thrombosis postoperatively because of their anti-platelet effect.

Sudden death may occur in CABG patients in the perioperative period because of spasm of a vein graft or particularly an internal mammary artery; 10 mg sublingual nifedipine is given and the intravenous nitroglycerin is increased. This event may require that the chest be opened in the I.C.U. or that 0.1-1.0 mg of nitroglycerin may be given directly down the bypass graft in the electrographic distribution of the ischemia. Coronary artery spasm may be a very important cause of postoperative sudden cardiac arrest for unexplained reasons. Nifedipine in these situations, if the patient survives, should be continued indefinitely postoperative. Internal mammary spasm has been shown to occur many months after the revascularization on occasion.

With regard to PTCA after prior coronary revascularization, PTCA of a vein graft is only 50% effective and is generally more effective closer to the distal anastomosis. Overall although PTCA is useful in patients previously undergoing coronary bypass, it is more useful in patients never having received CABG. Angiography generally underestimates the severity of atherosclerosis. In general, any vein graft greater than 5 years old should be replaced at the time of redo-coronary artery surgery.

TECHNIQUE OF CORONARY ARTERY BYPASS

The patient is brought to the operating room and an arterial line and Swan-Ganz catheter are placed for hemodynamic monitoring. The patient is induced under general endotracheal anesthesia. A Foley catheter is placed and the chest and legs are prepped and draped. A median sternotomy is performed. If the mammary needs to be taken, this is done with an internal mammary artery retractor. The vein is generally harvested from the right lower extremity during this time. The pericardium is opened. Heparin is given and pursestrings applied. The aorta is cannulated. The right atrial appendage is cannulated with a two-stage venous cannula with one port going into the inferior vena cava and the proximal port in the right atrium. An antegrade cardioplegia cannula/aortic root vent is placed in the aortic root, and a retrograde cannula is placed transatrially into the coronary sinus. Cardiopulmonary bypass is instituted, the aortic cross-clamp applied, and antegrade blood cardioplegia instilled through the aortic root until arrest occurs, then this is switched to retrograde. The chosen targets are then grafted with the saphenous vein graft. This includes grafting to the distal right coronary, posterior descending, posterolateral branches or acute marginal for the right coronary system. For the left anterior descending system, this may include branches of the diagonal as well. For the circumflex system, this may include branches of the obtuse marginal, as well as posterior ventricular branches in the case of a left dominant system.

The proximal coronary anastomosis onto the aortic root is performed either during the period of cross-clamping or once the cross-clamp is released with an aortic side-biter clamp. Once the aortic cross-clamp comes off, 100 mg of lidocaine is given to help limit the amount of arrhythmias. When the patient has gained a normal sinus rhythm, the bleeding has been controlled and mixed venous saturation is adequate, weaning is instituted. At this point, the anesthesiologist ventilates the patient. One gram of calcium is given by the perfusionist and the following drugs are administered to help limit the chance of a protamine reaction: 100 mg of solu-medrol, 300 mg of cimetidine, and 25 mg of benadryl. With weaning from bypass accomplished, the venous line is removed and protamine is given. Generally, 3 mg/kg of protamine is given (i.e., equivalent to the 3 mg/kg of heparin dose administered at the beginning of the case). The ACT which was initially greater than 400 prior to initiating bypass now comes down to approximately baseline, which is about 100 seconds.

In the event that the patient has difficulty weaning from bypass, then cardiopulmonary bypass is reinstituted and the inotropes need to be raised. Initially, weaning from cardiopulmonary bypass is performed with 1 mcg/kg/min of nitroglycerin and between 2.5 and 5 mcg/kg/min of dopamine. This may need to be increased to higher levels of dopamine and/or epinephrine added. If epinephrine needs to be added, this is usually started at about 1 mcg/min and can be titrated up. If this fails, then other pharmacologic maneuvers may be in order. If the patient has an elevated pulmonary artery pressure and also has a low cardiac output, this may be helped with Amrinone, which has a positive inotropic effect, but also causes vasodilatation, particularly on the pulmonary circuit. The choice of drug used depends on the hemodynamic parameters, the systemic vascular resistance, the cardiac output, and the pulmonary artery pressures.

In the event weaning is still unsuccessful, an intra-aortic balloon pump may be necessary. If weaning is still unsuccessful, i.e., after two tries at weaning with elevated inotropic agents and a balloon pump in place, then additional mechanical support is warranted. This type of support depends on the findings. If there is elevated pulmonary wedge pressure yet a low cardiac index (less than 1.5 l/min/m^2) and a low blood pressure (i.e., less than 90 mmHg), then it could be assumed that there is left ventricular failure and a LVAD is warranted. Conversely, if the patient has right ventricular failure, there will be elevated right atrial pressure and low left-sided filling pressures, as well as low cardiac output and low blood pressure, i.e., an inability to volume load the left heart despite an elevated right atrial pressure. In this event, an RVAD may be required.

The LVAD is a partial bypass circuit from the left atrium to a biomedicus vortex pump which then pumps the blood into the aorta. This bypass circuit can be used to achieve nearly full flow and requires only mild to moderate heparinization at approximately an ACT level of 150 seconds for flows below two liters. The LVAD is positioned in the left atrium via the right superior pulmonary vein via a pledget pursestring placed in the right superior pulmonary vein. The aortic cannula, already in place in the aorta, can be used as the outflow for the LVAD. An RVAD is placed via the right atrial cannula in place which then drains blood through the biomedicus vortex pump into a cannula positioned in the pulmonary artery. Placement of the LVAD can be life-saving, and several patients in our unit have successfully survived placement of the LVAD. Its use is basically to support a stunned heart which will eventually recover.

An additional note should be made on the use of the intra-aortic balloon pump. As stated previously, this is useful to decrease afterload, thus decreasing the workload on the heart and increases coronary artery perfusion pressure by inflating in diastole. There are some patients who cannot have an intra-aortic balloon pump. Such patients have extremely atherosclerotic aorta-iliac disease which precludes placement of even a guidewire, let alone a balloon pump into their aorta. In these patients, an alternative route can be placement of the balloon pump directly into the ascending aorta, and then threading the balloon pump distally past the left subclavian into the descending aorta. This is done through pursestrings placed as for the aortic root cannula.

There may be vascular complications related to the balloon pump. These include lower extremity ischemia and necrosis. It is important to document lower extremity pulses preoperatively and, of course, to continue to follow pulses closely postoperatively when a balloon pump is positioned. If there is evidence of ischemia, then a decision must be made whether the patient can tolerate weaning from the balloon pump or if the balloon pump must be replaced to the contralateral extremity. Alternatively, a femoral-femoral bypass with 8 mm Dacron graft may be necessary.

Redo-coronary artery surgery is much more hazardous than first time coronary surgery. An oscillating saw is used to enter the sternum. Caution must be used to avoid injuring the heart. When dissecting the heart from surrounding tissues, caution must be taken that the old vein grafts are not excessively manipulated because they may release emboli into the heart resulting in a trash heart which can result in ischemia and infarction.

In general, after cardiopulmonary bypass is instituted, retrograde cardioplegia is very useful to flush atheromatous emboli out of the old vein grafts. These can be transected, the distal limb tied off, and the proximal limb used for the proximal anastomoses of the new vein grafts. Because of the increased risk of atheromatous emboli going down patent old grafts, these anastomoses are done first. After giving the initial dose of retrograde cardioplegia the old grafts are transected distally and proximally and the distal anastomosis is performed followed by the proximal anastomosis. Since it may be difficult to place a side-biter clamp for construction of proximal anastomosis with the cross-clamp off, the proximal anastomoses are generally performed during the period of cardiac arrest.

LEFT VENTRICULAR ANEURYSM

Left ventricular aneurysm is, on occasion, an indication for surgery. The indications for operating on a left ventricular aneurysm are large size with decrease in cardiac output and congestive heart failure and predisposition to ventricular arrhythmias. Ventricular aneurysms are seen to have a paradoxical outward motion in ventricular systole (i.e., the ventricle is dyskinetic rather than akinetic). The dyskinetic ventricle results in decrease in ejection fraction by sequestering the cardiac output within the paradoxically moving aneurysm sac. The dilated aneurysm causes an increase in wall stress by the law of La Place, resulting in increased myocardial oxygen utilization and worsening ischemia. This may result in congestive heart failure. The aneurysm and surrounding scar tissue predispose to arrhythmias and this too is an indication for surgery. Ventricular aneurysms are usually located anterolaterally, although 20% of the time they may be located posteriorly (Fig. 5.1). The posterior defects are much more dangerous, and half of these are false aneurysms, i.e., the aneurysms have ruptured and have been contained. The mortality of operating on posterior aneurysms is much higher than that of anterolateral aneurysms. Cardiopulmonary bypass is instituted and the

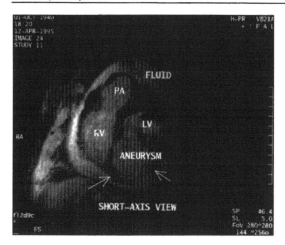

Fig. 5.1. Ultrafast CT scan demonstrating ruptured post-infarction ventricular aneurysm with fluid within pericardial sac.

5

aneurysm is incised after cross-clamping and cardioplegia solution has been given. The aneurysm is resected (Fig. 5.2a,b) and its margins can be noted by the smooth margin of the dead tissue compared to the rough trabecular surface of viable myocardium. Organized thrombus is frequently evident within the aneurysm (Fig. 5.2c). Coronary bypass is performed in standard fashion. The ventricular resection is repaired with felt strips and 3-0 or 4-0 Prolene horizontal mattress sutures. De-airing maneuvers are done as previously described (10 cm pressure on the lungs to fill the left heart; aspiration of the LA and LV with a needle and syringe). Only then is the cross-clamp released.

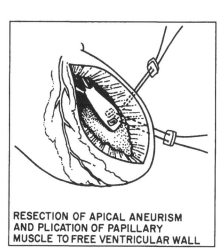

RESECTION OF APICAL ANEURISM AND PLICATION OF PAPILLARY MUSCLE TO FREE VENTRICULAR WALL

Fig. 5.2a. Resection of apical aneurysm and resuspension of papillary muscle to free ventricular wall.

Fig. 5.2b. Ventriculotomy closure after aneurysmectomy and coronary bypass using felt strips.

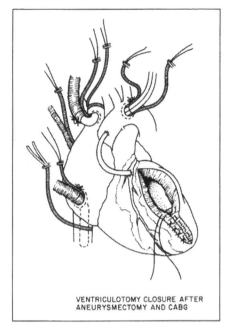

VENTRICULOTOMY CLOSURE AFTER
ANEURYSMECTOMY AND CABG

Fig. 5.2c. Organized thrombus extracted from wall of ventricular aneurysm during repair.

ISCHEMIC VENTRICULAR SEPTAL DEFECT
AND MITRAL REGURGITATION

POST-INFARCTION VENTRICULOSEPTAL DEFECT

Post-infarction VSD occurs through necrotic myocardium resulting from infarction and complicates in 1% of cases of acute myocardial infarction and usually occurs within 2 weeks following infarction. It is a true surgical emergency. About 25% of patients die within the first day and more than half die within the first week. After an acute myocardial infarction, if a patient develops a new systolic murmur or sudden congestive heart failure within 1-2 weeks, one should always be suspicious of either a post-infarction ventricular septal defect or post-infarction papillary muscle rupture. Either of these may result in the same type of symptoms and may be difficult to differentiate clinically because they both result in systolic murmurs and congestive failure. A Swan-Ganz catheter is extremely useful in differentiating between these since an oxygen saturation step-up in the pulmonary artery is consistent with a post-infarction VSD but not papillary muscle rupture resulting in mitral regurgitation. The left-to-right shunt resulting from a post-infarction VSD is what leads to the pulmonary artery oxygen step-up.

After the patient has been resuscitated and a diagnosis inferred from Swan-Ganz catheterization, the patient is brought to cardiac catheterization and cineangiography. This confirms the diagnosis of either post-infarction VSD or papillary muscle rupture and helps assess myocardial function. Also, it allows one to see if there is a concomitant ventricular aneurysm which requires repair. Cineangiography is essential as a road map to determine which vessels should be bypassed and where at the time of definitive surgery.

The morphology of post-infarction VSDs is most commonly anterior or apical. About 20% of patients have a VSD in the posterior portion of ventricular septum. Posterior VSD is frequently associated with mitral valve regurgitation secondary to papillary muscle infarction. It should be noted that ventricular aneurysm may be associated with post-infarction VSD.

The conduct of the operation includes bicaval cannulation with tapes. Cardiopulmonary bypass is instituted and after cold cardioplegic arrest, the VSD is approached through the left ventricle. In apical ventriculoseptal defects, the apical septum is approached via the anterolateral infarct that is almost always present. There may be an aneurysm associated with the infarction, in which case the incision is made directly through the aneurysm. The apical septal defect is repaired with a Dacron patch with pledgets on the right ventricular side and the suture is then brought up through the left ventricular side through the Dacron patch (Fig. 5.3a). If it is an apical infarction, one can amputate the apex to include the VSD and close with felt to include the septum and thus plicate the left ventricular free wall, the interventricular septum and the right ventricular free wall (See Fig. 5.3b).

For repair of a posterior septal VSD, the exposure is more difficult. This is done through the necrotic posterior left ventricular infarct. Pledgets are placed on the right ventricular aspect of the interventricular septum and attached to a Dacron patch located on the left ventricular side of the interventricular septum (Fig. 5.4).

In the case of posterior post-infarction VSD, there may be concomitant mitral regurgitation secondary to ischemia of the posterior medial papillary muscle. A patch may then be used to close the left ventricular infarction after it has been excised if necessary. It should be noted that operation early after rupture has a risk of death of about 50% compared to a risk of death of only 6% if the operation is done more than 3 weeks after rupture. One possible explanation for this is because the edges of the defect hold sutures better when an operation is delayed. Hence, if hemodynamically stable, some suggest delaying surgery, but usually the patient must go to surgery urgently because of the severity of their heart failure and hemodynamic compromise.

Fig. 5.3a. Anterior post-infarction VSD repaired by excising infarcted, aneurysmal tissue, performing a Dacron patch repair of the VSD, and felt strip closure of the ventriculotomy.

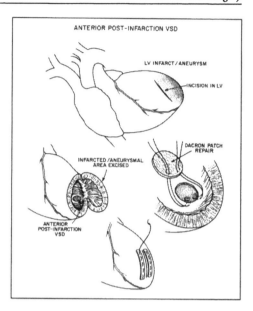

ISCHEMIC MITRAL REGURGITATION INCLUDING
POST-INFARCTION PAPILLARY MUSCLE RUPTURE

As mentioned in the previous section, it may be clinically impossible to differentiate between post-infarction VSD and post-infarction papillary muscle rupture resulting in mitral regurgitation. Both present with systolic murmurs and congestive heart failure and require urgent surgery.

If the patient presents with severe congestive heart failure and coronary artery disease, the patient is resuscitated and is sent to the ICU for Swan-Ganz catheterization and measurement of left- and right-sided heart pressures. Saturation of

Fig. 5.3b. Apical post-infarction VSD repaired by resection of the entire apex of the heart, including the VSD and portions of the left and right ventricles.

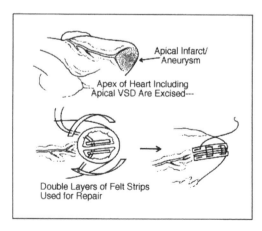

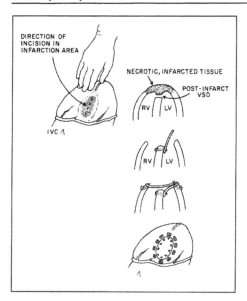

DIRECTION OF INCISION IN INFARCTION AREA

NECROTIC, INFARCTED TISSUE

POST-INFARCT VSD

RV LV

IVC Λ

RV LV

Λ

Fig. 5.4. Repair of posteriorly located post-infarction VSD requiring excision of nonviable tissue, plication of the interventricular septum to the free ventricular wall and closure of the right ventricle as a patch repair.

the right atrium and pulmonary artery are assessed to look for a step-up in the pulmonary artery saturation indicating a post-infarction VSD. If this is not found and if the patient does have a systolic murmur, then the patient may well have severe ischemic mitral regurgitation based on Swan-Ganz catheter readings.

If the patient is hemodynamically unstable, pharmacological support may be indicated and if refractory, an intra-aortic balloon pump may be inserted. It should be noted that the balloon pump may be extremely useful to hemodynamically support the patient because it will decrease afterload, thus limiting the amount of mitral regurgitation. This is similar to the beneficial effect of an intra-aortic balloon pump for post-ventriculoseptal defect in which case the decrease in afterload will limit the amount of left-to-right shunt. Expeditious CABG and MVR should be done since the mortality with expectant management is about 80%. In general, ischemic mitral regurgitation can be classified as acute or chronic. The acute variety is usually the result of papillary muscle rupture (Fig. 5.5) and the chronic variety is usually from ischemic elongation of the chordae or papillary muscles. Mitral regurgitation of ischemic etiology may be periodic and changes as the left ventricular function changes. Of patients with acute myocardial infarction, 1% die from severe mitral regurgitation from total papillary rupture. The posteromedial papillary muscle is more vulnerable to this phenomenon since it is supplied only by the posterior descending artery whereas the anterolateral muscle which is supplied by both the circumflex and LAD. Eighty percent of the time, ischemic mitral regurgitation is from the posterolateral papillary muscles; 20% of the time is from the anterolateral papillary muscle. Only about 20% of patients with acute papillary muscle rupture survive more than 24 hours if treated nonsurgically. Partial rupture has a much better prognosis. It should be noted that

Fig. 5.5. Ischemic ruptured papillary muscle attached to resected mitral valve leaflet. Courtesy of Dr. Lawrence Burr, University of British Columbia.

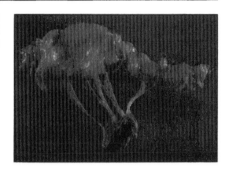

even with appropriate surgery, only two-thirds of patients with acute papillary muscle rupture survive.

How does one decide if a mitral valve replacement is needed in ischemic mitral regurgitation? Generally it depends on the severity of the mitral regurgitation and if there is organic disease involved. Obviously if there is organic disease of the valve itself as with rheumatic fever, valve replacement is indicated. If, however, the mitral regurgitation is only moderate based on the cardiac catheterization and echocardiogram, then mitral replacement or repair may not be indicated. Severe mitral regurgitation however, i.e., grade 3 out of 3, warrants inspection of the mitral valve with either repair or replacement. The mitral valve inspection and management is done after the distal anastomoses are performed and prior to performing the proximal anastomoses although the exact sequence is not critical. At this time, an annuloplasty ring or mitral valve replacement or other form of valve repair is done.

If ruptured chordae to the posterior leaflet are found, a quadrangular excision and annuloplasty may be done. However, because of the high mortality associated with ischemic mitral valve procedures and coronary artery bypass grafting, it is often recommended to simply do a valve replacement and coronary revascularization rather than attempt a repair that may not work. Therefore if doubt exists, a valve replacement should be done as a definitive operation rather than a valve repair. If doubt exists as to whether the patient needs any valve procedure done at all, i.e., if there is moderate mitral regurgitation presumably due to ischemia, then one can perform coronary bypass, then come off bypass and evaluate the patient's hemodynamics. Generally, if the cardiac output is adequate, the left atrial pressure parallels the right atrial pressure and there is no huge V-wave, then one can assume that there is not a large amount of mitral regurgitation and the patient does not need mitral valve replacement. One can also evaluate this with an echocardiogram intraoperatively via the transesophageal route. It is important to remember that combined coronary artery bypass graft and mitral valve replacement for ischemia has a higher mortality (15%), compared to combined coronary bypass with mitral valve replacement for other reasons (rheumatic heart disease or mitral prolapse from degeneration) which is in the range of 7%.

All patients with acute papillary rupture after acute myocardial infarction should have prompt Swan-Ganz catheter placement, cardiac catheterization and operation since they may decompensate acutely despite initial success with medical management. The indications for surgery with chronic papillary dysfunction from ischemia are less clear. The incompetence may fluctuate in degree. The decision on CABG, with or without mitral valve replacement or repair, depends on the degree of regurgitation as noted above. Again, in general, mitral valve repair should be avoided in acute severe mitral regurgitation, or if the anterior leaflet of the mitral valve is involved. It is important to remember that mitral regurgitation from ischemia may be due to annular dilation from the left ventricular failure in addition to ischemic papillary muscle dysfunction. A rule of thumb is that in patients with angina with mild to moderate mitral regurgitation and normal left ventricular ejection fraction, CABG alone is adequate. For angina with intermittent mild to moderate mitral regurgitation occurring only with ischemic attacks, coronary artery bypass grafting alone is surely adequate. In patients with severe mitral regurgitation with good left ventricular ejection fraction, a coronary artery bypass graft with mitral valve replacement or repair is indicated. In patients with severe mitral regurgitation and poor left ventricular ejection fraction, again mitral valve replacement may be done but one is less likely to take a chance with a mitral valve repair.

COMBINED CAROTID AND CORONARY ARTERY DISEASE

This is a complex and controversial topic. Several guidelines must be drawn up for indications for coronary artery bypass alone, compared to carotid endarterectomy alone. Combined carotid and coronary artery disease is common, particularly in patients with left main coronary artery disease. The chance of left main coronary artery disease in a patient who has combined carotid and coronary artery disease is one in three whereas patients without carotid disease but who do have coronary artery disease have a 14% chance of having left main disease. Conversely, if a patient has left main coronary artery disease, the chance of carotid disease is 30-50%; this is compared to a 20% incidence of carotid disease in patients with coronary artery disease without left main disease. *The point is that carotid artery disease must always be suspected in patients with coronary artery disease and in particular in patients who have left main coronary artery disease since one-third to half of these people will have concomitant, significant carotid disease.* If there is a neck bruit, whether or not it is symptomatic, a noninvasive evaluation, i.e., carotid duplex, is mandatory. In asymptomatic patients who have greater than 80% carotid stenosis, the chance of developing symptoms or developing more significant carotid lesions within 2 years is about 50%. It is therefore felt by most surgeons that carotid lesions greater than 80%, even if they are asymptomatic, should be treated by carotid endarterectomy. In general, the rule when deciding on the operation for combined carotid and coronary artery disease is that one operates first on the more compelling lesion; if both lesions are compelling then a combined procedure is done, i.e., a carotid endarterectomy is performed while the saphenous vein for coronary bypass is being harvested; the patient is then placed on cardiopulmonary bypass and the coronary artery bypass graft is done with the neck open. Then both wounds are closed after protamine is given to prevent excess bleeding from the neck wound.

Here are some examples: (1) The first is a patient with unstable angina, left main coronary artery disease and an asymptomatic 80% carotid stenosis. In this situation, coronary artery bypass is clearly the more compelling of the operations. This should be performed with the patient systemically cooled to protect the brain. The perfusion pressure is kept high. Later, after the patient has recovered from his coronary revascularization, a carotid endarterectomy should be performed because the chance of him developing a stroke, TIA or worsening carotid pathology within 2 years is about 50%. (2) The patient with stable angina, who has TIAs and a critical carotid lesion should clearly undergo a carotid endarterectomy first. After the patient has recovered, a coronary artery bypass graft is performed. Obviously, during the carotid endarterectomy, appropriate hemodynamic monitoring including a Swan-Ganz and arterial line may be critical. (3) In a patient who has unstable angina from left main coronary artery disease, as well as a critical carotid lesion with TIAs, a simultaneous procedure is performed in the fashion described above with the carotid endarterectomy done off bypass. For patients with unstable angina with critical coronary artery disease and a bilateral critical carotid lesion, then the surgeon can fix one carotid while off bypass, then perform coronary artery bypass grafting at the same sitting. Later, after the patient has recovered from his combined carotid-CABG, surgery, the surgeon can fix the other carotid, i.e., the less critical side. These examples serve as acceptable methods for approaching combined carotid and coronary artery disease; yet other approaches may work as well.

ISCHEMIC AND NONISCHEMIC VENTRICULAR TACHYCARDIA

The treatment of ventricular tachycardia (VT) during or after myocardial infarction varies depending upon the interval since the event. If the tachycardia occurs within 24-48 h of myocardial infarction, it is considered less malignant, especially if not sustained (less than 30 sec in duration). This often is caused by reperfusion of infarcted tissue and is not considered a malignant rhythm.

Runs of VT less than 30 sec that impair hemodynamics enough to cause symptoms of reduced blood flow are considered the equivalent of a sustained VT. These are generally considered to be included among the life-threatening cardiac arrhythmias. In the patient with prior myocardial infarction and a defined ventricular aneurysm, sustained VT occurs at rates from 140-200/min, and this arrhythmia may be relatively well-tolerated. In contrast, patients with transient myocardial ischemia often have more rapid ventricular tachyarrhythmias (>200/min) that may be polymorphic causing hemodynamic instability. When the hemodynamic status is stable and there is no evidence of myocardial ischemia, acute infarction or poor central nervous system perfusion, electrical cardioversion can await a therapeutic trial of intravenous drug. With acute myocardial infarction, falling blood pressure or evidence of ischemia, immediate cardioversion is indicated. If acute ischemia is not thought to be the cause, long-term therapy, with an automatic internal defibrillator or antiarrhythmic therapy depends upon the result of an electrophysiologic study as well as the severity of the clinical presentation. If ischemia is found to be the underlying cause of VT, vascularization is indicated with continued observation.

SUGGESTED READING

1. Karp RB. Special considerations in mitral valve and coronary artery disease. Adv Card Surg 1990; 1:71-92.
2. Lytle BW, Cosgrove DM, Gill CC et al. Mitral valve replacement combined with myocardial revascularization. Early and late results for 300 patients. Circulation 1985; 71:1179-1190.
3. Disesa VJ, Cohn LH, Collins JJ et al. Determinations of operative survival following combined mitral valve replacement and coronary revascularization. Ann Thorac Surg 1982, 34:482-489.
4. Minami K, Sagoo KS, Breymann J et al. Operative strategy in combined coronary and carotid artery disease. J Thorac Cardiovasc Surg 1988; 95:303-309.
5. Newman DC, Hicks RG. Combined carotid and coronary artery surgery: A review of the literature. Ann Thorac Surg 1988; 45:574-581.
6. Foster AH, Salter DR. Cerebral pathophysiologic considerations in patients with co-existing carotid and coronary artery disease. Adv Card Surg 1991; 2:203-225.
7. Coulden R, Lipton MJ. Noninvasive therapy for the diagnosis and management of myocardial ischemia. Adv Card Surg 1993; 4:173-205.
8. Elefferiades JA, Biblo LA, Batsford WP et al. Evolving patterns in the surgical treatment of malignant ventricular tachyarrhytmias. Ann Thorac Surg 1990; 49:94-100.
9. Klein H, Trappe HJ. Implantable cardioverter-defibrillator therapy. Indications and decision making in patients with coronary artery disease. PACE 1992; 15:610-615.
10. Dagget WM, Buckley MJ, Akins CW et al. Improved results of surgical management of post-infarction ventricular septal rupture. Ann Surg 1981; 196:269.
11. Heitmiller R, Jacobs ML, Daggett WM. Surgical management of post-infarction ventricular septal rupture. Ann Thorac Surg 1986; 41:683.
12. Silva JP, Cascudo MM, Baumgratz JF et al. Post-infarction ventricular septal defect: An efficacious technique for early surgical repair. J Thorac Cardiovasc Surg 1989; 97:86.
13. Piwrica A, Menasche P, Beaufils P et al. Long-term results of emergency surgery for post-infarction ventricular septal defect. Ann Thorac Surg 1987; 44:274.
14. European Coronary Surgery Study Group. Long-term results of prospective randomized study of coronary artery bypass surgery in stable angina pectoris. Lancet 1982; 2:1173-1180.
15. CASS principal investigators and their associates. Coronary artery surgery study (CASS): A randomized trial of coronary artery bypass surgery. Survival data. Circulation 1981; 68:939-950.
16. The Veterans Administration Coronary Artery Bypass Cooperative Study Group, 1984. Eleven year survival in the Veterans Administration Randomized Trial of Coronary Bypass Surgery for stable angina. N Engl J Med 1984; 311:133-139.

17. Passamani E, Davis KB, Gillespie MJ et al. A randomized trial of coronary artery bypass surgery. Survival of patients with a low ejection fraction. N Engl J Med 1985; 312:1665-1671.

18. Kouchoukos NJ, Ksrp RB, Oberman A et al. Long-term patency of saphenous veins for coronary bypass grafting. Circulation 1978; 58(Suppl I):96.

19. Lytle BW, Loop FD, Cosgrove DM et al. Long-term serial studies of internal mammary artery and saphenous vein coronary bypass grafts. J Thorac Cardiovasc Surg 1985; 89:248.

20. Loop FD, Lytle BW, Cosgrove DM et al. Influences of the internal mammary artery graft on 10-year survival and other cardiac events. N Engl J Med 1986; 314:1-6.

21. Cameron A, Kemp HG, Green GE. Bypass surgery with the internal mammary artery graft: 15 year follow-up. Circulation 1986; 74 (Suppl III); 30-36.

22. Jatene AD. Left ventricular aneurysmectomy. Resection of reconstruction. J Thorac Cardiovasc Surg 1985; 89:331.

23. Barratt-Boyes BG, White HD, Agnew JM et al. The results of surgical treatment of left ventricular aneurysms. An assessment of the risk factors affecting early and late mortality. J Thorac Cardiovasc Surg 1984; 87:87.

24. Olearchyk AS, Leomole GM, Spagna PM. Left ventricular aneurysm. Ten years experience in surgical treatment of 244 cases. Improved clinical status, hemodynamics, and long-term longevity. J Thorac Cardiovasc Surg 1984; 88:544.

Valvular Heart Disease

AORTIC VALVE DISEASE

ANATOMY

The aortic valve is a tricuspid valve with a right, left and noncoronary cusp (Fig. 6.1). The right coronary orifice is in the sinus of Valsalva of the right coronary cusp; the left main coronary ostium is in the sinus of Valsalva of the left coronary cusp. The junction between the left and right coronary cusps is the intercoronary commissure; the junction between the right coronary cusp and the noncoronary cusp is the anterior commissure; the junction between the noncoronary cusp and the left coronary cusp is the posterior commissure.

As noted earlier, the membranous septum is just underneath the junction of the right coronary cusp and noncoronary cusp and immediately beneath this lies the bundle of His. This is an area which requires particular attention during surgery to prevent injury resulting in heart block. Subaortic extension of calcification of the aortic valve may lead to heart block because of involvement of this area.

PATHOLOGY

Aortic valve disease can result in aortic stenosis or aortic insufficiency. The most common cause of aortic stenosis is calcific aortic stenosis in a congenitally bicuspid aortic valve (Fig. 6.2). Nearly as common is senile calcific aortic stenosis with severe calcification secondary to age. The third important cause of aortic stenosis is rheumatic fever. Aortic insufficiency has a more varied etiology. Rheumatic disease is the most common cause of aortic insufficiency. Unlike rheumatic aortic stenosis, commissural fusion is not present in rheumatic insufficiency and the leaflets are only minimally thickened. With rheumatic aortic stenosis, the leaflets are markedly thickened. The basic pathology of aortic insufficiency related to rheumatic disease is related to shortening of the cusps.

Other causes of aortic insufficiency include bicuspid or unicuspid aortic valves, endocarditis, and annular ectasia from a chronic aortic aneurysm or from acute dissection. The normal area of the aortic valve is 2-3 cm^2. An area less than 0.8 cm^2 can be considered severe stenosis, as is a gradient across the aortic valve of greater than 50 mm mean gradient.

The indications for surgery for aortic stenosis include the classic symptoms of syncope, angina or heart failure. Of these, angina is the most common symptom

Fig. 6.1. Aortic valve anatomy.

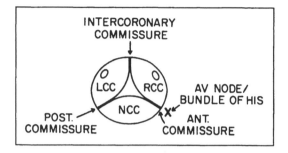

Fig. 6.2. Severe calcification in a congenitally biscuspid aortic valve with stenosis. Courtesy of Dr. Lawrence Burr, University of British Columbia.

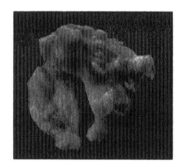

occurring in half of patients requiring aortic valve replacement for aortic stenosis. All three symptoms are present in about a third of patients presenting with symptomatic aortic stenosis. These symptoms usually exist when the aortic valve gradient is greater than 50 mmHg or the area of the valve is less than 1 cm².

Indications for operating on patients with aortic insufficiency are much more complex and less clear-cut than for stenosis. Generally, Class III or Class IV heart failure status is an indication for surgery. For Class I or Class II New York Heart Association classification, surgery is only performed if there are signs of left ventricular dysfunction; for example, if the left ventricular ejection fraction is less than 40% or if the ventricular function worsens after a stress test. The ejection fraction while undergoing a stress test is the most important parameter in evaluating whether to operate in aortic insufficiency (or mitral regurgitation for that matter). As an example, if a patient can increase his ejection fraction from 40 to 50% with exercise, then even though his ejection fraction is moderately diminished and the patient has Class I-II symptoms, the valve probably does not need to be replaced. If, however, he is unable to increase his ejection fraction with exercise, this is a sign of impending cardiac demise and a valve replacement is indicated. Other parameters that support aortic valve replacement in patients with Class I or Class II symptoms are left ventricular enlargement with a cardiac/thoracic ratio of greater than 0.55, or left ventricular enlargement with left ventricular end diastolic dimension of greater than 50 mm. It should be noted that aortic insufficiency is a very insidious disease. This is because it is a volume overload problem of the left ventricle, rather than a pressure overload problem as with aortic stenosis.

Aortic insufficiency progresses insidiously with even severe regurgitation not causing the symptoms in a patient until there is marked myocardial dysfunction. Once the patient starts to develop myocardial dysfunction and diminished ejection fraction, one can be assured that the aortic insufficiency has progressed far and that surgery is needed. There may be a point where the ejection fraction may be so bad with aortic insufficiency that aortic valve replacement will lead to the patient's demise. This is not true for aortic stenosis and one should consider aortic valve replacement in all patients with severe aortic stenosis, no matter how bad the ejection fraction because these patients will almost certainly do better after the aortic valve replacement. This may not be true for the longstanding ventricular dysfunction associated with aortic insufficiency.

For aortic insufficiency associated with annular ectasia from chronic aneurysm or acute dissection, replacement of the ascending aorta may be required as well as valve replacement. A composite graft replacement of the ascending aorta with an aortic valve replacement and reimplantation of the coronary arteries into the graft is a Bentall procedure. The details will be described in the section on Aortic Aneurysm and Dissection. In brief, cardiopulmonary bypass is established with an arterial line either in the aorta or femoral artery for the bypass. The aorta is cross-clamped and opened and cardioplegia solution infused into the coronary orifices directly or alternatively retrograde cardioplegia solution is infused. A Dacron graft which has a mechanical valve sewn into it by the manufacturer is used. The aortic valve replacement is performed first, followed by reimplantation of the coronary orifices, followed by sewing the distal anastomosis.

TECHNIQUE OF AORTIC VALVE REPLACEMENT (Fig. 6.3)
A standard median sternotomy is performed. The aorta is cannulated and then the right atrium is cannulated with a two-stage cannula drawing blood from both the inferior vena cava and right atrium. An antegrade plegia cannula is placed followed by a retrograde plegia cannula in the coronary sinus. Cardiopulmonary bypass is instituted. A left ventricular vent is placed via the right superior pulmonary vein. Cross-clamp is placed and antegrade cold blood cardioplegia given initially, then switched to retrograde cold blood cardioplegia. If the patient has severe aortic insufficiency, then no antegrade plegia is given and retrograde plegia is used from the start. This is given in an intermittent fashion since it is cold blood and there is no need for continuous plegia administration. Every 20 minutes, more plegia solution is given. With the heart arrested, the aortic root is opened approximately 2 cm proximal to the annulus in an oblique fashion. The aortic valve is excised, the annulus sized and pledgetted 2-0 Ethibond sutures are placed from the ventricle to the aortic side, except at the commissures at which point the stitches are placed from the aorta into the ventricular side to approximate the level of the annulus. This seats the valve in the supra-annular position.

If a suture breaks when placing the aortic valve it is not a simple matter to place another stitch into the annulus of the aorta and up into the sewing ring of the valve because the pledgets are placed from the ventricular up into the aortic aspect. If a suture does break, several techniques are available. First of all, the pledget

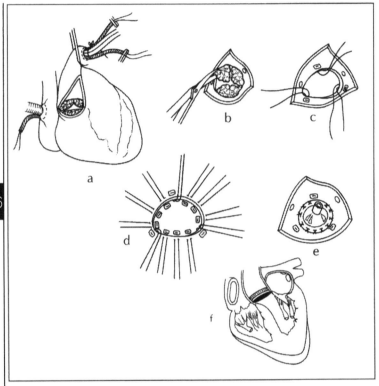

Fig. 6.3. Aortic valve replacement. a) Exposure of the aortic valve through an oblique aortotomy. b) Resection of the native aortic valve. The CUSA ultrasonic dissector may be very effective for managing a heavily calcified annulus. c) Commissural sutures are placed from the aortic into the ventricular direction. d) The remaining sutures are placed from the ventricular into the aortic direction. e) The sutures are placed through the prosthetic sewing ring and the valve seated and sutures tied. f) The final position of the prosthetic (St. Jude) aortic valve. Note that the valve position is supra-annular and the valve leaflets are perpendicular to the septum.

must be removed because it is free floating in the ventricle and could embolize. Sometimes, however, it may be wedged between the valve sewing ring and the annulus and therefore it may not be possible to retrieve the pledget. If the stitch breaks in the region of the noncoronary cusp, it will be possible to place a pledgetted suture from outside of the aorta, through the aorta and then up through the sewing ring of the valve. If the stitch breaks in the region of the left coronary cusp or right coronary cusp, one may not be so fortunate because placing such a pledget stitch in these cusps may result in tacking down the right or left coronary artery. This is why the sutures are initially tied down first in the right coronary cusp then the left coronary cusp prior to tying down the noncoronary cusp sutures. If a right coronary cusp stitch breaks or a left coronary cusp stitch breaks, then the valve can be lifted and it is a simple matter to place another stitch. If the right coronary cusp sutures are tied followed by the left coronary cusp sutures and then a break occurs when tying down the noncoronary cusp stitches, one can still salvage the situation by driving a pledgetted stitch from the outside of the aorta up through the aortic valve sewing ring. Each move done in cardiac surgery should have a function based on a knowledge of efficiency for the best possible outcome for the patient.

The aorta is closed, and prior to completing the closure the left ventricular vent is stopped and 10 cm pressure applied to the lungs to expel air. Then higher suction is placed on the left ventricular vent and aortic root vent is applied to further evacuate air. Then the cross-clamp is removed. The left ventricular vent is once again shut off and needle aspiration of the dome of the left atrium and the left ventricle performed. With the rhythm obtained and bleeding controlled, the left ventricular vent is once again shut off and removed and the patient weaned from cardiopulmonary bypass.

SMALL AORTIC ANNULUS

The management of a small aortic annulus may be difficult. A homograft may be used with minimal gradient (down to 16 mm size), but if a prosthetic must be used (i.e., St. Jude's mechanical) then the valve diameter in general should be greater than or equal to 21 mm at least, lest there be a large gradient across the valve with inadequate cardiac output during exercise. This judgment depends on the size of the patient and whether or not a higher cardiac output would be necessary, i.e., such as an athlete versus a sedentary patient. St. Jude's aortic valves of 17 mm size are available and are generally unacceptable in this size except for small or sedentary patients. As a rule of thumb, patients with a body surface area (BSA) less than 1.7 will manage with a 19 mm valve. Those between 1.7-2.0 m^2 will do best with a 21 mm valve, and those greater than 2.0 m^2 generally need a 23 mm valve.

The St. Jude HP® (Hemodynamic Plus) valve achieves a larger orifice for a given valve size (see section on Mechanical and Bioprosthetic Cardiac Valves). The result is an effective increase in the valve orifice area by nearly one valve size (i.e., a St. Jude 19 mm HP valve has nearly the orifice size of a standard St. Jude 21 mm valve). The HP valve is not recommended for the heavily calcified valve annulus, although we have used the valve safely after annular decalcification using the ultrasonic dissector (CUSA). The use of the HP valve markedly reduces the requirement for surgical enlargement of the aortic annulus.

Nonetheless, other surgical techniques may be necessary to seat a valve greater than or equal to 21 mm:

1. Simply placing the sutures from the direction of the ventricle up into the aorta with the pledget seated on the ventricular aspect allows for one valve size larger. This is because the sewing ring of the aortic valve thus sits in a supra-annular position. If the pledgetted sutures were placed from the aorta down into the ventricle, the valve would be seated intra-annular. At Harbor, we typically place the stitches such that the valve seats in the supra-annular position. This is also found to be highly effective for the HP valve.

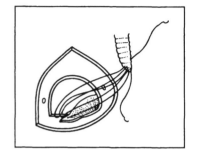

Fig. 6.4. Aortic annular enlargement. Posterior commissure between the left and noncoronary cusps is split down to the anterior leaflet of the mitral valve and a Dacron patch is used to enlarge the annulus. Generally, this enables the next larger valve size to be placed.

2. Supra-annular and annular patch enlargement can be performed by incising down the junction between the noncoronary and left coronary cusp (Manouguian procedure). By splitting this posterior commissure, one is able to then sew in an albumin-treated Dacron patch to enlarge the annulus (Fig. 6.4). Pledgetted sutures are then passed through the Dacron in standard fashion, as if there were an annulus present there, and the valve can be seated.

3. A less ideal method is the supra-annular patch enlargement by incising down into the noncoronary sinus but not actually into the annulus. This merely enlarges the supra-annular portion but not the annulus itself.

4. A Konno aortoventriculoplasty may also be performed. This is a procedure best done by pediatric cardiac surgeons and consists of a vertical aortotomy followed by incision just to the right of the commissure between the left and right coronary cusps. The right ventricle is opened with an oblique incision and the interventricular septum is then incised, thus splaying open the annulus. A piece of Dacron is sewn into the interventricular septum and up into the annulus and then used to cover the aorta. The right ventriculotomy is covered with a pericardial patch. This is rarely needed if ever.

5. Another method primarily of historic interest is placement of a valved conduit from the left ventricular apex to abdominal aorta.

MITRAL VALVE DISEASE

ANATOMY

The mitral valve is a bi-leaflet structure and as described before the surface area of the anterior leaflet is much greater than that of the posterior leaflet, although the posterior leaflet extends along more of the circumference of the annulus than the anterior leaflet (Fig. 6.5).

The left fibrous trigone has the greatest risk of structural injury to the aortic valve in the region of the noncoronary cusp and left coronary cusp. The right

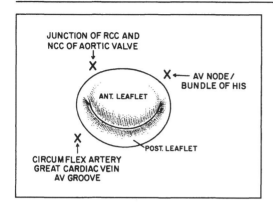

Fig. 6.5. Mitral valve and neighboring structures.

6

fibrous trigone is in the region of the AV node and bundle of His. At approximately the 7 to 8 o'clock position on the mitral annulus is the region of the AV groove and circumflex coronary artery and coronary sinus. It is at this point that the greatest risk to these structures exists, and disruption at this point may lead to catastrophic AV disruption leading to an actual separation of the left atria and left ventricle, i.e., separation of the back wall of the heart itself.

PATHOLOGY

Mitral valve disease can be classified as mitral stenosis or mitral regurgitation. Mitral stenosis is nearly always rheumatic in etiology although functionally it may result from such things as a large left atrial myxoma plugging up the mitral valve in a ball-valve type action. Mitral stenosis may lead to increased pulmonary vascular resistance due to spasm of the pulmonary arteries as well as back pressure directly from the mitral valve to the left atrium and then to the pulmonary veins. The elevated pulmonary vascular resistance may lead to right heart failure or tricuspid regurgitation.

Mitral regurgitation, like aortic regurgitation, is due to a variety of causes. The most common cause of mitral regurgitation is mitral valve prolapse. This occurs from myxomatous valve degeneration or from localized idiopathic chordal rupture resulting from localized myxomatous degeneration (usually in the posteromedial portion of the posterior leaflet).

The next most common cause of mitral regurgitation is rheumatic mitral regurgitation. If a patient has combined mitral stenosis and mitral regurgitation, the cause is nearly always rheumatic. Rheumatic mitral regurgitation results from two possible etiologies. One is the acute rheumatic process resulting in annular dilation from myocarditis. This occurs from the actual rheumatic infection. However, the most common cause is from the chronic sequelae of rheumatic fever resulting in severe valve thickening, especially involving the posterior leaflet, as well as rolling of the leaflet edges. The chordae may be elongated but more typically are shortened, thickened and fused, and sometimes they are so short that the papillary muscles actually appear to be fused up to the valve leaflet. When describing rheumatic mitral valve leaflets, one must assess the commissural fusion, leaflet thickening, leaflet calcification, leaflet fibrosis and chordal shortening. Other causes of mitral regurgitation are ischemic mitral regurgitation which has been extensively described previously. Endocarditis is another etiology of mitral regurgitation.

The normal mitral valve area is 4-6 cm². Symptoms typically start when the area is less than 2 cm². If less than 1 cm², the patient is generally symptomatic at rest (New York Heart Association Class IV). The indications for operation for mitral stenosis is New York Heart Association Class III or IV failure. These patients generally require a mitral valve replacement. New York Heart Association Class II patients generally require commissurotomy but do not necessarily require a valve replacement. Generally, a patient who has a mitral valve area less than 1.5 cm² or a gradient across the mitral valve of greater than 15 mm requires some mitral valve procedure, be it a commissurotomy or a mitral valve replacement. For mitral regurgitation, the indications again are more vague than those for mitral stenosis. The indications for mitral regurgitation are the same as those for aortic insufficiency. Like aortic insufficiency, mitral regurgitation is a condition of volume overload of the ventricle. New York Heart Association Class III or IV is an indication for mitral valve replacement for mitral regurgitation. Those New York Heart Association Class I or II patients are generally not operated on for mitral regurgitation unless they have signs of impending heart failure and have diminution of ventricular function indices such as diminished left ventricular ejection fraction of less than 40% or an ejection fraction which worsens after a stress test. Left ventricular enlargement with an enlarged cardiac to thoracic ratio is an indication for surgery as it is in aortic insufficiency. Again, once a patient develops signs of failure with mitral regurgitation, the disease has been longstanding and the changes may not be reversible.

An interesting phenomenon that occurs postoperatively in patients who undergo mitral valve replacement for mitral regurgitation is that their ejection fraction may actually worsen. This is because of the increased afterload after mitral valve replacement. In other words, one has eliminated the path of least resistance from the ventricle back up into the atrium and now the ventricle can only eject out of the aorta. This results in a higher left ventricular filling pressure on the Starling's curve and may result in a reduced ejection fraction (Fig. 6.6). There is some evidence that preservation of the chordae maintains left ventricular geometry and prevents this increase in left ventricular and diastolic filling pressure, thus keeping the patient on a favorable portion of the Starling curve. Hence, chordal preservation is recommended in cases of mitral valve replacement for mitral regurgitation.

After aortic valve replacement for aortic insufficiency, conversely ejection fraction can improve as left ventricular filling pressure decreases because of relief of the aortic insufficiency. Therefore one may move to a more favorable position on the Starling's curve.

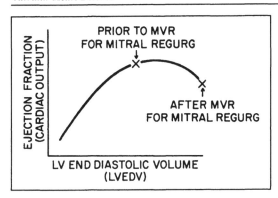

Fig. 6.6. Starling curve depicting a shift of myocardial performance to a less favorable position after mitral valve replacement for mitral regurgitation. The decompression of the left ventricle into the left atrium has been eliminated, increasing the LVEDV and worsening the ejection fraction.

TECHNIQUE OF MITRAL VALVE REPLACEMENT

A standard median sternotomy incision is performed. The aorta is cannulated. Separate cannulas are placed in the superior vena cava and inferior vena cava, i.e., bicaval cannulation, because this permits increased mobility of the left atrium which improves exposure of the mitral valve. The most common difficulty associated with mitral valve replacement is inadequate exposure of the valve. Techniques that can be used to improve visualization of the mitral valve include incision along the superior vena cava and inferior vena cava to mobilize these; separate cannulation of the superior and inferior vena cava; mobilization of the right superior pulmonary vein to prevent tethering down of the left atrium; and extensive mobilization of the undersurface of the rectus fascia and inferior portion of the wound to improve the opening into the chest.

After the aorta and cavae are cannulated, an antegrade plegia cannula is placed followed by a retrograde plegia cannula through the right atrium into the coronary sinus. After cardiopulmonary bypass is instituted, 10 cm pressure is applied to the lungs and a left atrial vent is placed via the right superior pulmonary vein and put on active suction. The aorta is cross-clamped and antegrade plegia is used to arrest the heart followed by retrograde cold blood cardioplegia. The left atrial vent is turned off and an incision is made in the left atrium alongside the groove just anterior to the right superior pulmonary vein. This groove corresponds to the interatrial septum (Fig. 6.7a).

The incision in the left atrium is extended proximally underneath the superior vena cava and distally underneath the inferior vena cava. Another approach is the transatrial, transseptal approach of Dubost (Fig. 6.7b). A mitral valve retractor is positioned and held by the first assistant. A traction stitch is placed in the anterior leaflet and mitral valve and pulled down to improve exposure of the mitral valve (Fig. 6.8). A knife is used to incise the base of the anterior mitral leaflet just sufficiently to place 2-0 pledgetted Ethibon stitch in the annulus at 12 o'clock to put downward traction on the annulus to improve exposure. This is a critical maneuver in cases of difficult mitral valve exposure and makes a sometimes im-

Fig. 6.7a. Mitral valve exposure. Standard mitral exposure. Left atriotomy begins just anterior to right superior pulmonary vein. This affords excellent exposure of the mitral valve in most cases.

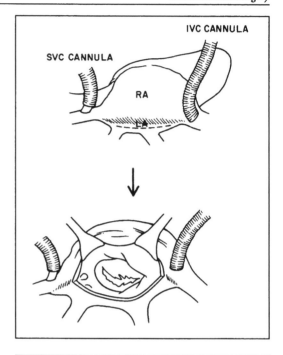

Fig. 6.7b. Dubost mitral exposure. A bi-atrial, transseptal incision is made going through the interatrial septum at the fossa ovalis. Bicaval cannulation and snares are required.

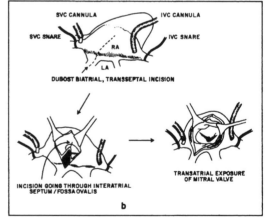

possible task fairly easy. The pledgetted Ethibond stitches are placed from the atrial into the ventricular aspect as the anterior leaflet is serially excised. This permits seating of the valve eventually in the intra-annular position rather than the supra-annular position, differing from aortic valve replacement. This is the recommended valve seating arrangement suggested by St. Jude's corporation, since the intra-annular position of the valve prevents the cordae and annulus from impinging

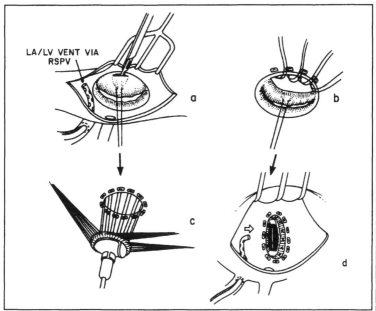

Fig. 6.8. Mitral valve replacement. a) Traction is placed on the anterior leaflet of the mitral valve to gain exposure of the 12 o'clock position. b) Annular sutures are placed sequentially, starting at the 12 o'clock position. Traction on this stitch will improve visualization of the annulus. Plegetted sutures are placed from atrial to ventricular aspect. c) Sutures are placed through the valve sewing ring and the valve seated. d) Seated mitral prosthesis. The vent is passed through the valve into the left ventricle.

on the leaflets of the mechanical valve. Furthermore, if a suture breaks, the pledget can be easily retrieved. (Some, however, support seating the valve in the supra-annular position, i.e., by placing the pledgetted sutures from the ventricular up into the atrial side. This results in a sturdy seating arrangement resulting in a higher burst pressure).

Excision of the valve continues down to the commissures and in general the posterior leaflet of the mitral valve is preserved so as to retain the chordae and preserve left ventricular geometry. If this cannot be done because of heavy calcification or fibrosis, then the posterior leaflet too should be excised. The sutures are then placed into the sewing ring of the St. Jude's valve. The leaflets should be positioned perpendicular to the interventricular septum as this allows for more clearance from the leaflets. Figure 6.9 shows that the clearance from the leaflet edge is greater when the valve is seated perpendicular to the annulus rather than parallel to the annulus.

It should be noted that for a porcine or bovine bioprosthesis, the stitches are likewise placed from the atrial to the ventricular side. The stents in the case of bioprostheses are positioned at the 10 o'clock to 2 o'clock position to prevent impingement of the left ventricular outflow tract. The widest cusp on the porcine bioprosthesis is placed adjacent to the septum (Fig. 6.10).

During the entire process of mitral valve replacement, care is taken to prevent injury of the circumflex and coronary sinus, AV node/bundle of His, and the noncoronary cusp left coronary cusp.

With the mitral valve seated in place, a left atrial vent is then placed across the mitral valve leaflets into the left ventricle, and the left atrium is then closed. Prior to finishing the left atrial closure, the left ventricular vent is shut off and 10 cm of pressure applied to the lungs to purge air. Then once the left atrium is closed, the left ventricular vent is restarted as is the aortic root vent, and the aortic cross-clamp is released with the patient in Trendelenburg. A needle and syringe are used to aspirate air from the dome of the left atrium and left ventricular apex, and weaning from coronary artery bypass is achieved as previously described.

Several technical points are worthy of mention. The surgeon stands on the

Fig. 6.9. Clearance of St. Jude valve leaflets. a) Minimal clearance of (1) (i.e., leaflets oriented perpendicular to septum) is greater than (2) (i.e., parallel to septum). b) Correct orientation of St. Jude valves in aortic and mitral positions with leaflets perpendicular to septum.

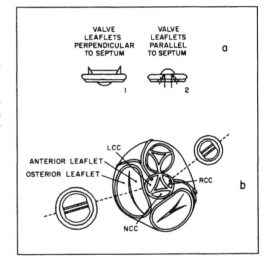

Fig. 6.10. Technical detail of suture orientation for porcine tissue valve. The widest prosthetic cusp ((A) to (B)) is oriented from (A) to (B) on the mitral annulus, corresponding to the left ventricular outflow tract under the aortic valve.

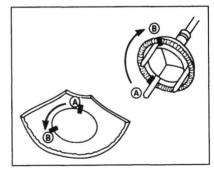

right hand side for this, as with all cardiac surgical procedures. The first assistant initially holds the mitral valve retractor. Once the valve is excised and the suture is placed, the first assistant gives the mitral valve retractor to a nurse or second assistant who stands on the left side of the first assistant. The third assistant stands on the right side of the primary surgeon. Thus the first assistant has both hands free and with his left hand holds the valve, and with his right hand holds a needle driver. The surgeon drives the sutures through the sewing ring and the first assistant grasps the needles in the sewing ring and pulls them out and inserts them into the fingers of his left hand which is holding the valve. At each one-third of the way around the circumference of the valve, the threads are clamped and then cut by the second and third assistants. The third assistant on the surgeon's right hand side loads the needle which is given to him by the surgeon, (i.e., the second half of each pair) in a backhand manner. An organized consistent approach by the surgeon, first, second and third assistants and scrub nurse makes the procedure run smoothly.

APPROACHES TO THE MITRAL VALVE

The standard approach is an incision just posterior to the inter-atrial groove into the left atrium, anterior to the right superior pulmonary vein (Fig. 6.7a). This approach is acceptable in nearly all cases as long as there has been adequate mobilization of the right superior pulmonary vein, superior and inferior vena cava and traction stitches have been placed in the anterior leaflet of the mitral valve to bring the annulus into view. If these maneuvers fail then other techniques are described but rarely needed. This includes the Dubost bi-atrial, transseptal approach to the mitral valve which includes an incision into the right atrium, left atrium, and through the inter-atrial septum to gain improved exposure of the mitral valve (Fig. 6.7b). Another approach is through the dome of the left atrium. Yet another approach is a left thoracotomy to gain access to the left atrial appendage with cannulation of the femoral artery and placement of another cannula in the right ventricular outflow tract into the right ventricle. The mitral valve can also be approached via a left ventricular aneurysm and replacing the valve from the ventricular aspect. Another approach is a right thoracotomy with standard cannulation giving improved exposure of the left atrium. Another method is through transsection of the superior vena cava and retraction inferior and laterally of the cava to gain access to the dome of the left atrium. Mitral valve replacement has also been approached through the aortic root via the aortic valve.

ATRIOVENTRICULAR DISRUPTION

Atrioventricular groove disruption is a horrifying complication which may occur after mitral valve replacement. This occurs because of a separation between the atria and the ventricle in the back of the heart. It may be due to a heavily calcified mitral valve annulus which cracks when the mitral valve is placed, or it may be due to heavily pulling on the heart to de-air the apex of the left ventricle. This axial torsion on the heart may actually tear the atrioventricular septum at the point that the ridge of the mitral valve attaches to it. This is exactly in the region of the circumflex coronary artery and coronary sinus, complicating things further. If a sudden gush of blood comes from the back of the heart after mitral valve replacement, one must assume an AV disruption. Although the chance of survival is limited at this point, several rapid maneuvers may be helpful in trying to salvage the patient. Cardiopulmonary pulmonary bypass must be reinstituted immediately. A cross-clamp is applied and plegia given. The left atrium is opened and the prosthetic valve is removed. Pledgetted sutures are used to repair the AV disruption from both within and outside the heart. Figure 6.11 shows the sutures placed on the outside of the heart. One must be extremely careful to avoid injury to the circumflex coronary artery and coronary sinus. Atrioventricular stability may be extremely tenuous. An episode of hypertension or other acute increase in afterload may result in a blow-out of a tentative atrioventricular groove.

MITRAL VALVE REPAIR

Mitral valve repair is becoming more and more common because of the advantages inherent in repairing the patient's own native mitral valve. It prevents the coagulation disorders necessary with coumadinization for mechanical valves and prevents the necessity for reoperation for bioprosthetic valve failure. The indications for mitral valve repair for mitral regurgitation include ruptured chordae to the posterior leaflet usually from myxomatous degeneration, or less commonly ischemia. Another indication is ischemic cardiomyopathy causing dilation of the mitral valve ring, in which case an annular ring (Duran ring) is placed. Another reason is myxomatous degeneration resulting in dilation of the annulus and redundancy of the leaflets. Another type of mitral valve repair is commissurotomy in the case of commissural fusion. This is only worthwhile if the leaflets themselves are pliable and not very thickened and the leaflet edges are intact without being rolled or deformed. The chordae and papillary muscles must be relatively normal. Chordal rupture to the anterior leaflet may also be repaired in some instances by chordal transfer from the posterior leaflet to the anterior leaflet (via

Fig. 6.11. Repair of atrioventricular rupture following mitral valve replacement. Plegetted sutures are positioned around the circumflex artery and coronary sinus.

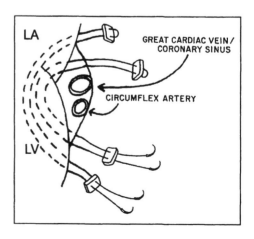

LA

GREAT CARDIAC VEIN / CORONARY SINUS

CIRCUMFLEX ARTERY

LV

quadrangular excision of the posterior leaflet and repair). It should be said that mitral valve repair, with the exception of commissurotomy, should always be accompanied by an annuloplasty since this supports the repair and will help prevent future dilation. Generally, a Duran ring, which is a complete ring, is used rather than a Carpentier ring which is an incomplete ring. The Carpentier ring is more rigid than the flexible Duran ring and is not as mobile with the annular movements of the normal cardiac cycle.

It is very important when doing the initial assessment of the mitral valve prior to repair to do a careful evaluation of the valve. This involves assessing the amount of leaflet fibrosis, commissural fusion, leaflet edge rolling, and chordae and papillary muscles, since this will determine whether repair is possible and the type of repair. Figure 6.12a-d describes the technical considerations of several types of mitral valve repairs.

One of the complications of mitral valve repair includes systolic anterior motion (SAM) of the anterior leaflet of the mitral valve. This can lead to left ventricular outflow obstruction from displacement of the anterior leaflet toward the septum with systole. Besides left ventricular outflow obstruction, SAM can also lead to severe mitral regurgitation. The phenomenon may occur when there is redundant tissue in a prolapsing mitral valve. After repair and annuloplasty, a redundant posterior leaflet fills the annuloplasty ring and results in coaptation of the posterior leaflet near the middle of the anterior leaflet rather than on its free edge. This shifts the anterior leaflet toward the ventricular septum, obstructing left ventricular outflow and distorting normal coaptation leading to mitral regurgitation.

SAM worsens with ionotropic support whereas beta blockade may be helpful. Afterload reduction, as with the use of IABP, also worsens SAM. Surgical maneuvers to treat SAM include myomectomy and septal resection although this entails additional risk and does not treat the underlying problem. Direct approaches in-

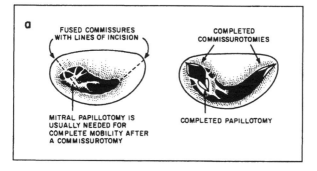

Fig. 6.12a. Mitral valve repair. Systematic valve evaluation includes assessing the leaflets, commissures, chordae and papillary muscles. a) Commissurotomy. Fused commissures are incised. Papillary muscle incision is frequently needed for complete mobility after a commissurotomy.

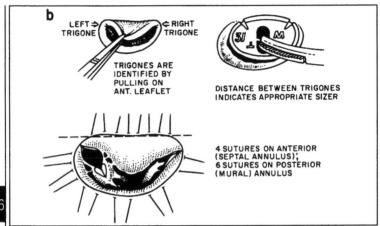

Fig. 6.12b. Annuloplasty. The trigones are identified by traction on the anterior leaflet. The distance between the trigones indicates the appropriate Duran ring size. Four nonplegetted sutures are placed on the anterior (septal) annulus. Six sutures are placed on the posterior (mural) annulus. For the anterior 4 sutures, the intrasuture distance (A) and intersuture distance (B) is the same on the annulus and Duran ring. For the posterior 6 sutures, the intrasuture distance (A) is greater on the annulus than on the Duran ring. This will tighten a dilated annulus. The intersuture distance (B) is the same on the annulus and Duran ring.

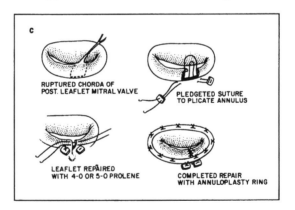

Fig. 6.12c. Quadrangular excision of the posterior leaflet of the mitral valve. Resection of the posterior leaflet containing the ruptured chordae is performed. Plegetted sutures are used to plicate the annulus. The leaflet is repaired with interrupted 5-0 Prolene. A Duran annuloplasty ring is used to reinforce the repair.

clude replacement of the mitral valve with a prosthesis. Another method to prevent or treat SAM is to reduce the height of the posterior leaflet in valve repair. This can be done by a sliding repair to detach and reposition a portion of the

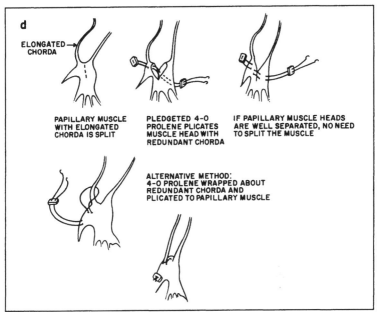

Fig. 6.12d. Chordal shortening. Plegetted 4-0 Prolene is used to plicate the muscle head with the redundant chordae. Alternately, a suture is used to wrap the redundant chorda and plicate it to the papillary muscle.

posterior leaflet to shorten its height. The resulting coaptation of the mitral leaflets is nearer to the leaflet edges, preventing posterior displacement of the anterior leaflet.

TRICUSPID VALVE DISEASE

ANATOMY

The tricuspid valve is a trileaflet structure consisting of anterior, posterior and septal leaflets, the anterior leaflet being the largest and posterior leaflet being the smallest (Fig. 6.13). The tricuspid valve leaflets have primary, secondary and tertiary chordae and papillary muscles. There is a single large anterior papillary muscle attached to the free wall and several accessory papillary muscles attached to the septum. The septal leaflet is fixed. The AV node and bundle of His run through the triangle of Koch, the borders of which are defined by the tendon of Todaro, the annulus of the tricuspid valve and the coronary sinus. The closest region between the annulus and the conduction tissue is where the bundle of His lies adjacent to the anterior and septal portions of the tricuspid valve; this is the area in

Fig. 6.13. Tricuspid valve and its relationship to the AV node and bundle of His. Danger points for injury to AV node and bundle of His indicated by row of Xs.

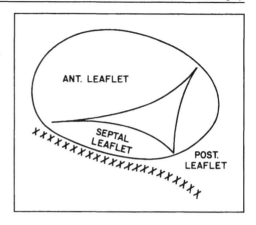

most danger of injury to the conduction tissue.

PATHOLOGY

Tricuspid valve disease is only rarely from tricuspid stenosis. When it is, it is nearly always of rheumatic etiology. Much more commonly, tricuspid disease is from tricuspid regurgitation. The most common type of tricuspid regurgitation is secondary tricuspid regurgitation, i.e., a functional disorder rather than an organic disorder of the tricuspid valve itself. This functional disorder of the tricuspid valve results from annular dilation as a result of elevated right ventricular pressure usually resulting from mitral valve disease or left heart failure. The functional dilation of the tricuspid usually involves dilation of the anterior two-thirds of the tricuspid annulus, especially in the region of the posterior leaflet. The septal leaflet is fixed between the trigones and does not dilate because it is part of the fibrous skeleton in the heart. In a similar fashion, the anterior leaflet of the mitral valve is fixed in position and usually does not dilate, whereas the posterior portion of the annulus mitral valve does dilate. The other type of tricuspid regurgitation is of rheumatic etiology.

When the tricuspid valve has rheumatic pathologic changes, the rheumatic process nearly always is associated with mitral and aortic valve disease as well. The tricuspid valve is the least common of these three valves to be involved with rheumatic disease. Rheumatic tricuspid regurgitation is usually associated with a variable amount of tricuspid stenosis. The hemodynamic effect of a moderate tricuspid stenosis is equivalent to a tight mitral stenosis because the right-sided pressures are low compared to left-sided pressure. Another etiology for tricuspid regurgitation is endocarditis. Intravenous drug abusers frequently develop vegetations on the tricuspid valve leading to valvular insufficiency.

When there is organic tricuspid pathology, as with rheumatic tricuspid disease or endocarditis, tricuspid valve surgery is indicated. Indications for operation on a functional tricuspid regurgitation are less clear. If a mitral valve replacement is being performed and the patient also has tricuspid valve regurgitation, then the indications for tricuspid valve replacement or repair are as follows: (1) The tricuspid regurgitation preoperatively has been important and constant, (i.e., not intermittent with periods of no tricuspid regurgitation during adequate medical management of the mitral disease); (2) Right ventricular dilation is longstanding; or (3) Pulmonary hypertension is evident. These would lead one to conclude that the likelihood of the functional tricuspid regurgitation resolving spontaneously after the mitral valve procedure is unlikely and that a tricuspid procedure is indicated.

Whether a replacement or repair of the tricuspid valve should be done depends on the status of the valve itself. If there is organic disease, generally replacement should be done. If the leaflets themselves are normal and the process is functional dilation, then repair can be performed. This can be one of several types: A Kay annuloplasty basically involves plication of the annulus in the region of the posterior leaflet (Fig. 6.14a). This narrows the annulus and basically obliterates the posterior leaflet of the tricuspid valve. The other procedure is annuloplasty of the tricuspid valve using a Duran ring (Fig. 6.14b). The size of the ring is estimated either by the diameter between the commissures of the anterior leaflet of the tricuspid valve, or by actually measuring the area of the anterior leaflet of the tricuspid valve and using this as the appropriate size for the Duran ring.

The DeVega annuloplasty is a tricuspid repair in which two limbs of a pledgetted stitch are brought around the annulus of the septal and anterior leaflet. The stitch is tightened such that only two gloved fingers may be passed through the tricuspid orifice. The conduction tissue adjacent to the septal annulus is totally avoided (Fig. 6.14c).

In tricuspid valve replacement, a portion of the valve leaflets are removed if

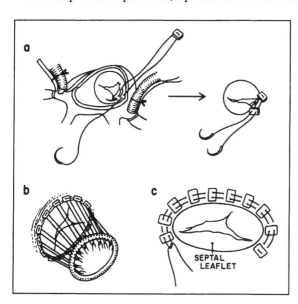

Fig. 6.14. Tricuspid valve repair. a) Kay plication annuloplasty of posterior leaflet. b) Duran tricuspid annulo-plasty ring. c) DeVega tricuspid annuloplasty. The septal leaflet is avoided.

SEPTAL LEAFLET

they are in the way, otherwise, they may be simply plicated out of the way using the annular sutures. It is important to remember that when using the sutures at the septal portion, the suture should not actually be placed into the annulus of the septal portion but rather within the base of the septal leaflet. This is to prevent injury to the conduction tissue. The base of the septal leaflet serves as a useful buttress and will hold the valve in position. Generally a Carpentier-Edwards bioprosthetic valve is used rather than a St. Jude's valve, although if a mitral valve replacement with a St. Jude's valve has been performed and the patient needs to be anticoagulated anyway, then there is some justification for placing a mechanical valve. However, this is suboptimal because a Swan-Ganz catheter cannot be placed through the St. Jude's valve in the tricuspid position but it can be when a Carpentier-Edward's bioprosthesis is placed. Triscuspid valve replacement is shown in Figure 6.15.

Technique of Tricuspid Valve Surgery

A median sternotomy is performed in the standard fashion, and the aorta is cannulated followed by bicaval cannulation in both the superior vena cava and inferior vena cava. Caval tapes are passed around the superior and inferior vena cava. Isolated bicaval cannulation is required to prevent air from entering the cavae with the right atrium open, lest an air lock develop which would shut off the pump machine. Also without the caval tapes, there would be blood rushing into the right atrium obscuring the operating field. After the aorta and cavae have been cannulated, an antegrade plegia line is placed (in case cardioplegia is needed), cardiopulmonary bypass is initiated and caval tapes snared. The right atrium is opened and surgery is done on the tricuspid in the open, beating heart. Remem-

Fig. 6.15. Tricuspid valve replacement. A Carpentier-Edwards tissue valve is seated into the tricuspid annulus.

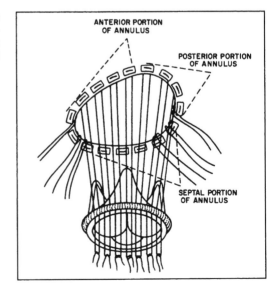

ber that surgery can be done on a beating heart on the right side only, and that the left side should never be opened while the heart is beating lest air be sucked into the left heart and then embolize out of the beating heart through the aorta. With the atrium open, the pump sucker is placed in the coronary sinus to drain blood coming out of the sinus and the tricuspid valve is inspected. If tricuspid valve replacement is required, the orifice is sized and the valve replaced. If the tricuspid valve needs repair only, then a Duran ring may be used with the size based on the area of the anterior leaflet. The right atrium is closed and caval tapes are loosened. Weaning from cardiopulmonary bypass is performed. If too much bleeding occurs during tricuspid valve procedures such that visualization is obscured, antegrade cold blood cardioplegia is given intermittently with the aorta cross-clamped. The valve procedure is done, then the aortic cross-clamp is released and the right atrium is closed with the cross-clamp off.

MECHANICAL AND BIOPROSTHETIC CARDIAC VALVES

A mechanical valve should be expected to last the lifetime of the patient. Bioprosthetic valves are made from porcine or bovine valve or pericardial tissue. They deteriorate with time, either becoming calcified or stenotic, or more commonly, the leaflet tears in the region of the stent post (Fig. 6.16). The advantage of the mechanical valve is that it is permanent. The disadvantage is that it requires anticoagulation for the rest of a patient's life. Even with adequate anticoagulation, there remains a chance of thrombosis.

Overall, there is a 5% chance of thromboembolism in 5 years with a mechanical prosthesis. There is about a 3% chance of developing an anticoagulation related complication in 5 years as a result of hemorrhage from the coumadin.

The advantage of the bioprosthetic valve is that there is no need for anticoagulation with no thrombotic or hemorrhagic complications. The disadvantage is that the valve deteriorates with time such that by 15 years, half of patients will require redo-cardiac surgery to replace the valves. Tissue valves deteriorate much faster in young patients. The reason for this is unclear, but it is felt to probably

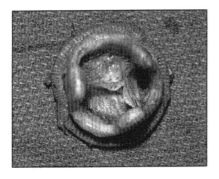

Fig. 6.16. Explanted Carpentier-Edwards bioprosthesis with a leaflet tear resulting in severe regurgitation.

represent differences in calcium metabolism in young people compared to older people. Also, the difference in force of contraction in younger people contributes to valve deterioration. In any event, in patient's under 50, a bioprosthesis deteriorates within 10-15 years in about half of patients and has an even worse record for younger patients.

It is interesting that tissue valves placed in the aortic position tend to deteriorate less rapidly than tissue valves placed in the mitral position. This is because when the ventricle contracts and there is high pressure within the ventricle, much greater stress is placed on the mitral valve than on the aortic valve, since during systole the aortic valve leaflets must simply open to permit flow, whereas the mitral leaflets are forcefully brought together leading to an increased rate of deterioration. For aortic valves, as a general rule, patients who are less than 70 years of age should undergo mechanical valve replacement because of the high chance that they will require reoperation at a later time if a tissue valve is placed. On the other hand, for patients who are greater than 70, a porcine or bovine bioprosthesis should be used. However for mitral valves, patients should be greater than 75 before a bioprosthesis is used because of the increased chance of deterioration of the mitral position. Patients less than 75 requiring mitral valve replacement generally should undergo placement of a mechanical prosthesis.

In younger people, exceptions may be made to the rule of using mechanical prostheses. For example, in an athlete who engages in high risk sports where there is a chance of head injury or hemorrhage, anticoagulation is a poor idea and a tissue valve should be placed, knowing full well that the patient will require another operation, perhaps quite early on. Also, a lady in childbearing age who wishes to have more children may be advised to have a tissue valve placed, again knowing that she will require reoperation for replacement of the tissue valve. If a patient with a mechanical valve does become pregnant, she should bear her children. The main problem with coumadin is the chance of teratogenicity. Therefore, coumadin should be stopped during the first trimester and the patient given heparin. Coumadin can be resumed the second trimester but then stopped again in the third trimester and the patient heparinized until the time of childbearing, at which time the heparin is stopped.

The mechanical valves that one can choose from include bi-leaflet and tilting disk mechanical valves (Fig. 6.17). St. Jude manufactures the valve with the most clinical experience and has basically remained unchanged since its inception over quarter-century ago. It is the valve used most commonly at our institution. The Medtronic-Hall, while not being in clinical use as long as the St. Jude's valve, is reported to be reliable as well. The Starr-Edwards valve was the original heart valve and works as a ball-in-cage type mechanism. It was altered to include a cloth covering over the struts which worsened the quality of the valve because cloth could embolize. This version is no longer being used. The Bjork-Shiley tilting disc and Duromedics bileaflet valves have both been pulled during their history because of the risk of fracture of the valve with embolizaton of the leaflets (Fig. 6.18). A modified Duromedics valve, however, has recently been introduced. Another valve, Carbomedics, is similar to the St. Jude's valve with the exception of a

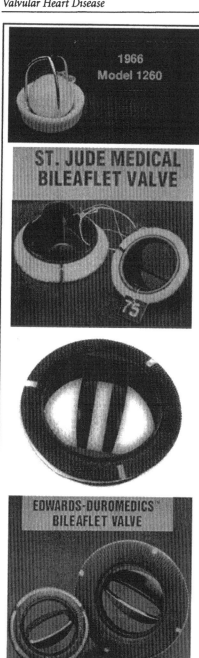

Fig. 6.17a. Mechanical prosthetic valves. Starr-Edwards Silastic Ball valve. (Courtesy Baxter Healthcare Corporation)

Fig. 6.17b. St. Jude bileaflet valve. (Courtesy Baxter Healthcare Corporation)

Fig. 6.17c. Carbomedics bileaflet valve.

Fig. 6.17d. Duromedics bileaflet valve. (Courtesy Baxter Healthcare Corporation)

Fig. 6.17e. Medtronic-Hall™ tilting disc valve.

Fig. 6.17f. Omniscience tilting disc valve.

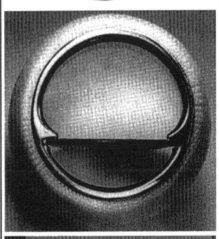

Fig. 6.17g. Bjork-Shiley tilting disc valve. (Courtesy Baxter Healthcare Corporation)

60° CONVEXO-CONCAVE
1976-1983

hinge modification and the fact that it can be rotated to allow for different orientations of the valve within the annulus once seated. Newer St. Jude's valves, however, also can rotate. Although used in other countries previously, it has only become available in the United States for about a decade. St. Jude's has developed a

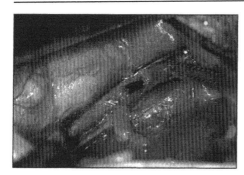

Fig. 6.18. Fracture embolization of a Duromedics mitral prosthetic leaflet to the iliac artery.

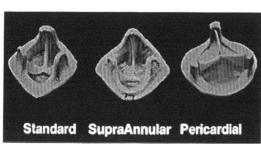

Fig. 6.19a. Bioprosthetic valves. a) Carpentier-Edwards standard and supra- annular porcine valves, and bovine pericardial valves. (Courtesy Baxter Healthcare Corporation)

Standard   SupraAnnular   Pericardial

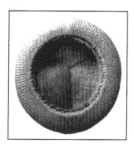

Fig. 6.19b. Medtronic Hancock® porcine valve.

new "HP" valve standing for "Hemodynamic Plus" valve, which permits a greater flow orifice for any given valve size. This is because the sewing ring is placed further up the annulus rather than in the annulus, meaning that less of the sewing ring is needed to seat the valve, thus the orifice can be bigger. Each size of the St. Jude's HP valve corresponds to one size up of the regular St. Jude's valve, i.e., a 19 mm HP St. Jude's valve corresponds to a 21 mm regular St. Jude's valve. This corresponds to approximately 25% greater flow through any given valve using the HP valve.

Tissue valves on the market include the Carpentier-Edwards porcine bioprosthesis and bovine pericardial bioprosthesis (Fig. 6.19a), the Hancock porcine valve (Fig. 6.19b), the Ionescue-Shiley bovine pericardial prosthesis, and the Medtronic Intact valve. The Edwards bovine pericardial aortic bioprosthesis appears to have superior long-term results to the porcine valve, with overall freedom of valve dysfunction at 91% over 12 years of study. It is now the most frequently implanted aortic bioprosthesis in the United States and is the tissue valve of choice at Harbor-UCLA. The mitral version of the pericardial valve is now on the market. The Intact valve uses nonpressure fixation of the valve, decreasing the chance of tearing by eliminating the collagen pattern responsible for making the leaflet stiff.

TECHNIQUES OF COMBINED VALVE PROCEDURES

Combined valve or valve-coronary procedures are commonplace. A systematic, routine approach is a requirement for optimizing outcomes. Our techniques at Harbor-UCLA are presented here.

COMBINED MITRAL VALVE REPLACEMENT AND AORTIC VALVE REPLACEMENT

Combined mitral and aortic valve replacement is performed using antegrade and retrograde intermittent cold blood cardioplegia. Although we have used continuous warm blood cardioplegia, we find that the aortic valve region is flooded from retrograde flow from the coronary orifices, making the procedure more difficult. Cardiopulmonary bypass is performed as previously described using bicaval cannulation. The aorta is cross-clamped, cardioplegia given antegrade and the heart arrested. The aorta is opened and the aortic valve removed. The left atrium is opened in standard fashion. A mitral valve replacement is done. The left atrial vent is then placed in the left ventricle and the left atriotomy closed. De-airing of the atrium is not necessary at this point as the aorta, still open, communicates with the left heart. After the left atrium has been closed, pledgetted sutures are applied to the aortic annulus and the annulus is sized. The conduct of the operation, therefore, is to perform the aortic valve resection first, then the mitral valve resection and replacement, and then the aortic valve replacement last. This is to prevent manipulation of a replaced aortic valve while trying to replace the mitral valve, which could lead to dehiscence of the aortic prosthesis from the annulus.

COMBINED MITRAL VALVE REPLACEMENT AND TRICUSPID VALVE REPLACEMENT

Intermittent antegrade and retrograde cold blood cardioplegia is administered and a left atrial vent is used. Cardiopulmonary bypass is initiated using bicaval cannulation with caval tapes. The cross-clamp is applied and the heart arrested using antegrade then retrograde cardioplegia. The left atrium is opened and a mitral valve replacement is performed. The left atrial vent is then placed in the left ventricle and the left atrium is closed. Prior to finishing the closure, the left ventricular vent is shut off and 10 cm pressure applied to the lungs for de-airing of the left heart. When the left atrium is closed, the LV vent is turned on again, the root vent is turned on, the aortic cross-clamp is released, and the patient is placed in Trendelenburg. The vents are shut off and air is aspirated from the dome of the left atrium and left ventricular apex. The right atrium is then opened and the tricuspid valve is replaced or repaired. Alternatively, one can do the tricuspid valve replacement in the arrested heart with the cross-clamp on to improve visualization, but in general it is better to have the cross-clamp off to have reperfusion of the heart.

COMBINED CORONARY BYPASS/VALVE REPLACEMENT

The conduct of the operation is to perform the distal anastomoses first with the coronaries attached to a turkey-foot perfusion cannulae for continuous infusion. Then the valve replacement is done, either aortic or mitral or both. The proximal anastomoses are next performed. Combined coronary/valve procedures are generally done using antegrade and retrograde cold blood cardioplegia in an intermittent fashion.

ENDOCARDITIS

Endocarditis of a cardiac valve can be one of the most devastating infections a physician must manage. Native valve endocarditis may occur in patients who are intravenous drug abusers; in these patients the tricuspid valve is usually involved. The aortic and mitral valves may be involved. Typically, intravenous drug abusers have *Staphylococcus aureus* as the primary organism. This generally represents acute bacterial endocarditis. Patients with valvular endocarditis who are not drug abusers may obtain their infections from dental abscesses, tooth extraction or oral surgery. The organism is frequently Streptococcal and represents subacute bacterial endocarditis.

Prosthetic valve endocarditis occurs in a post-surgical infection of a heart valve. Early on in the perioperative period and even up to 1 year, *Staph epidermidis* and gram-negative organisms are the most common. Late prosthetic valve endocarditis usually involve Streptococcal organisms. Patients who have native valve endocarditis may present in a number of ways. These include fever, chills, night sweats and signs of systemic bacteremia, or they may present with shortness of breath, heart failure and signs of hemodynamic compromise. Patients suspected of having endocarditis should undergo an echocardiogram which can be invaluable in assessing valve function and determining if there are vegetations attached to the valve. Cardiac catheterization and cineangiography may be indicated. The patient should be stabilized by afterload reduction in the case of aortic insufficiency and mitral insufficiency and admitted to the Intensive Care Unit if necessary. Blood cultures should be taken and the patient empirically treated with broad-spectrum antibiotics. The antibiotic regimen should be modified according to culture results.

The indications for surgery for native valve endocarditis are as follows:

1. Hemodynamic compromise with heart failure. This typically occurs as sudden and severe aortic or mitral regurgitation in a heart that has not been accustomed to chronic and slowly worsening regurgitation pattern.
2. Persistent bacteremia after at least 1 week of adequate antibiotic therapy or resistant organism.
3. Massive destruction of the valve with an annular abscess.
4. Recurrent embolization by the vegetation or a persisting vegetation which may embolize at any time.

It is important to perform a CT scan of the brain and spleen since optimally other septic organs are managed before the heart surgery, although this is not always feasible. In general, if at all possible, the patient should undergo at least 1 week of intravenous antibiotics prior to cardiac surgery intervention.

For prosthetic valve endocarditis, surgery has been shown to significantly improve survival compared to medical management. Prosthetic valve endocarditis usually involves a periprosthetic leak. For a periprosthetic infection, antibiotics are given preoperatively, then the valve is replaced. Replacement is particularly compelling if there is evidence of periprosthetic leak, septic embolization, or refractoriness of the infection. In selected patients, if there is good response to antibiotics and there is no periprosthetic leak or other complications, one may simply manage the infection with antibiotics.

In the case of a destroyed aortic annulus due to an abscess in either a native valve endocarditis or prosthetic valve endocarditis, one could do a Bentall type ascending aortic and aortic valve replacement with deep seating of the valve in the left ventricular outflow tract and reimplantation of the coronary arteries. Alternatively a homograft valve with its attached aorta may be deeply seated into the left ventricular outflow tract with reimplantation of the coronary arteries.

For tricuspid valve endocarditis a similar protocol may be followed as with the aortic and mitral valves with the exception that tricuspid valve excision alone without replacement may be a management option. Patients who present with tricuspid valve endocarditis once again should be treated with antibiotics, and if there is poor response to the antibiotic treatment or if there are other compelling reasons for surgery the patient may simply undergo tricuspid valve excision. The patient is carefully watched in the perioperative period. If there are signs of severe tricuspid regurgitation with ascites, massive hepatomegaly, lower extremity edema and inadequate hemodynamics because of poor volume loading of the left heart, then reoperation may be necessary and the tricuspid valve replaced with a bioprosthesis.

More and more surgeons appear to be performing primary tricuspid valve replacement rather than simple tricuspid valve excision as the initial surgery to prevent the frequent hemodynamic sequelae of tricuspid valve excision alone.

AORTIC ALLOGRAFT AND AUTOGRAFT PROCEDURES

Many cardiac surgeons are developing increased interest in the use of aortic allograft (homograft) and pulmonary autograft to manage a wide variety of aortic valve and root pathology. Allografts procured from organ donors are cryopreserved and commercially available in a variety of sizes. Because anticoagulation can be avoided and because of the low incidence of valve failure, thromboembolic events, and endocarditis, the use of allograft and native pulmonary tissue is particularly attractive. One of the most compelling indications for allograft procedures is in acute aortic endocarditis with massive tissue destruction resulting in aortic ventricular discontinuity. Being a biologic device, these alterna-

Fig. 6.20. a) Freehand (infracoronary) allograft. b) Allograft for complete root replacement with coronary ostia holes constructed. c) Inclusion cylinder technique.

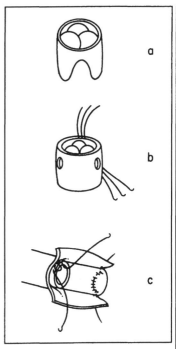

6

tive methods are not expected to surpass or equal the longevity of most mechanical aortic valves. There is increasing evidence that the longevity of homografts and autografts is superior to porcine bioprostheses, although the comparison to bovine pericardial bioprostheses is not complete at present.

Allograft aortic procedures can be summarized as freehand allograft (subcoronary) technique, complete root replacement, and inclusion cylinder technique. In the freehand allograft technique, the graft is trimmed to include only the valve leaflets and commissures. Three tissue "posts" are left remaining between which is left enough room to avoid obstructing the patient's native coronary ostia (Fig. 6.20a). The allograft is sewn into position in the aortic annulus and the tissue "posts" are sewn to the native aortic wall, avoiding the coronary ostia.

The most important factor in determining short and long term success for the freehand allograft technique is the precision of insertion. Slight technical errors which may not be apparent at the initial operation lead to geometric imperfection, distortion and increased failure rate. This has led many to abandon the freehand technique in place of complete allograft root replacement which eliminates errors of geometric misalignment (Fig. 6.20b). Short and long term studies clearly demonstrate that such complete root replacement procedures have lower pressure gradients and lower incidence of aortic insufficiency than the freehand technique. One technical modification is the inclusion cylinder technique, in which

Fig. 6.21a. Ross pulmonary autograft procedure. Aorta is opened and aortic valve removed.
Courtesy of Dr. Jeff Milliken, Harbor-UCLA.

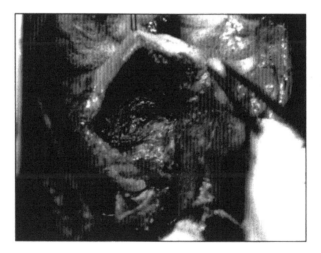

Fig. 6.21b. The pulmonary valve is excised en bloc with portions of the right ventricular outflow tract and pulmonary artery. A large residual defect in the right ventricle is left remaining.
Courtesy of Dr. Jeff Milliken, Harbor-UCLA.

the allograft root replacement is performed within the walls of the native aorta,
which is then wrapped around the allograft (Fig. 6.20c). Advantages of this technique include protection from suture line bleeding, but a disadvantage is that if

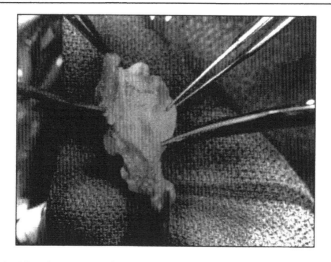

Fig. 6.21c. The pulmonary autograft is trimmed. Courtesy of Dr. Jeff Milliken, Harbor-UCLA.

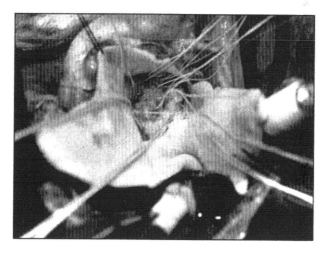

Fig. 6.21d. The pulmonary autograft is sutured to the aortic annulus. Courtesy of Dr. Jeff Milliken, Harbor-UCLA.

bleeding does occur between the wrap and the allograft, distortion of the valve leaflets or coronary ostia may occur.

The pulmonary autograft (Ross) procedure for aortic valve replacement involves replacing the patient's aortic valve and root with a pulmonary autograft, then using an allograft in the pulmonary position (Fig. 6.21a-f). The rationale for using such a complex procedure is the increased durability of the pulmonary autograft compared to an allograft in the aortic position. Additionally, the allograft

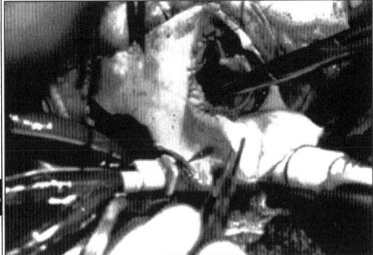

Fig. 6.21e. The distal anastomosis of the pulmonary autograft is performed. Courtesy of Dr. Jeff Milliken, Harbor-UCLA.

Fig. 6.21f. An aortic homograft is prepared for insertion into patient's pulmonary position. Courtesy of Dr. Jeff Milliken, Harbor-UCLA.

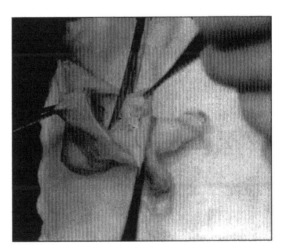

has increased durability in the pulmonary position compared to the aortic position. The procedure has been controversial because of its complexity for what is usually a straightforward operation and probably should only be performed in experienced centers.

SUGGESTED READING

1. Stapleton JF. The natural history of chronic valvular disease. Cardiovasc Clin 1986; 16(2):105-47

2. Dervan J, Goldberg S. Acute aortic regurgitation. Pathophysiology and management. In: Frankel WS, Brest AN, eds. Valvular Heart Disease: Comprehensive Evaluation and Management. Cardiovasc Clinics 1986; 16(2):281-288.

3. Gaasch WH. Aortic valve disease: timing of valve replacement surgery. In: Starek PJK, ed. Heart Valve Replacement and Reconstruction. Chicago: Year Book Medical Publishers, 1987:21-29.

4. Richardson JV, Karp RB, Kirklin JW et al. Treatment of infective endocarditis: A 10 year comparative analysis. Circulation 1978; 58:589.

5. Morgan RJ, Davis JT, Fraker TD. Current status of valve prosthesis. Surg Clin North Am 1985; 65:699.

6. Manouguian S, Seybold-Epting W. Patch enlargement of the aortic valve by extending the aortic incision into the anterior mitral leaflet: New operative technique. J Thorac Cardiovasc Surg 1979; 78:402-412.

7. Ross DN. Application of homografts in clinical surgery. J Cardiac Surg 1987; 1(Suppl):177-183.

8. Kouchoukos NT, Davila Roman VG et al. Replacement of the aortic root with a pulmonary autograft in children and young adults with aortic valve disease. N Engl J Med 1994; 330:1-6.

9. Frater RVM. Functional anatomy of the mitral valve. In: Ionescu MI, Cohn LH, eds. Mitral Valve Disease. London: Butterworths, 1985:127.

10. David TE, Uden DE, Strauss HD. The importance of the mitral apparatus in left ventricular function after correction of mitral regurgitation. Circulation 1983; 68 (Suppl 2); 76-82.

11. Davies MJ. Aetiology and pathology of the diseased mitral valve. In: Ionescu MI, Cohn LH, eds. Mitral Valve Disease. London: Butterworths, 1985:27-41.

12. Gorlin R. Natural history, medical therapy and indications for surgery in mitral valve disease. In: Ionescu MI, Cohn LH, eds. Mitral Valve Disease. London: Butterworths, 1985:105-123.

13. Yacoub M, Halim M, Radley Smith R et al. Surgical treatment of mitral regurgitation caused by floppy values; repair versus replacement. Circulation 1981; 64(Suppl 2):210.

14. Peterson KL, Tajimi T. The timing of surgical intervention in chronic mitral regurgitation. In: Duran C, Angell WW, Johnson AD et al, eds. Recent Progress in Mitral Valve Disease. London: Butterworths, 1984:171-180.

15. Chaux A, Matloff JM, Czer LSC et al. Clinical results with the St. Jude medical valve. In: Starch PJK, ed. Heart Valve Replacement and Reconstruction. Chicago: Year Book Medical 1987; 191-200

16. Pipello DF, Bessone LN, Blank RH et al. The porcine bioprosthesis: patient age as a factor predicting failure. In: Bodnar E, Yacoub M, eds. Biologic and Bioprosthetic Values. New York: Yorke Medical Books, 1986:130-134.

17. Rivera R, Duran E, Ajuria M. Carpentier's flexible ring versus De Vega's annuloplasty. J Thorac Cardiovasc Surg 1985; 89:196.

18. Jones EL, Shah VB, Shanewise JS et al. Should the freehand allograft be abandoned as a reliable alternative for aortic valve replacement? Ann Thorac Surg 1995; 59:1397-1404.

19.   Larbelestier RI, Chard RB, Cohn LH. Optimal approach to the mitral valve: Dissection of the interatrial groove. Ann Thorac Surg 1992; 54:1186-8.

20.   Horskotte D, Shulte HD, Bircks W et al. The effects of chordal preservation on late outcome after mitral valve replacement: A randomized study. J Heart Valve Dis 1993; 2:150-8.

21.   Hennein H, Swain JA, McIntosh CL et al. Comparative clinical assessment of mitral valve replacement with and without chordal preservation. J Thorac Cardiovasc Surg 1990; 99:828-37.

22.   Akins CW. Selection of cardiac valvular prostheses. Ann Thorac Surg 1993; 55:801-2.

23.   Akens CW, Hilgenberg AD, Buckley MJ et al. Mitral valve reconstruction versus replacement for degenerative or ischemic mitral regurgitation. Ann Thorac Surg 1994; 58:668.

24.   Enriquez-Sorano M, Schaff HV, Orzulak TA et al. Valve repair improves the outcome of surgery for mitral regurgitation: A multivariate analysis. Circulation 1995; 91:1022.

25.   Duram CMG. Tricuspid valve surgery revisited. J Cardiac Surg 1994; 9(Suppl):242-7.

26.   Cohn LH. Tricuspid regurgitation secondary to mitral valve disease: When and how to repair. J Cardiac Surg 1994; 9(Suppl):237-41.

27.   Prabhakar G, Kumar V, Gometza B et al. Surgery for organic rheumatic disease of the tricuspid valve. J Heart Valve Dis 1993; 2:561-6.

28.   Baumgartner FJ, Omari B, Robertson JM et al. Annular abscesses in surgicla endocarditis: anatomic, clinical and operative features. Ann Thorac Surg 2000; 70:442-7.

29.   Grossi EA, Steinberg BM, LeBoutillier M et al. Decreasing incidence of systolic anterior motion after mitral valve reconstruction. Circulation 1994; 90:(5Pt2)II:195-7.

30.   Jebara V, Mihaileanu S, Acar C et al. Left ventricular outflow tract obstruction after mitral valve repair: Results of the sliding leaflet technique. Circulation 1993; 88:(5Pt2)II:30-4.

Aortic Aneurysm/Dissections/Ruptures

Fritz J. Baumgartner

GENERAL CLASSIFICATION

The topic of aortic aneurysms, dissections and ruptures is a difficult one in terms of the classification and management. First, the common misconception that the thoracic aortic aneurysm is the same as a dissection should be dispelled. These are different entities and their management is different with one exception. When an aortic dissection has persisted for a long period of time, it becomes a chronic aortic dissection which should actually be classified under the term chronic aortic aneurysm. There is no such thing as an acute aortic aneurysm. There is, however, such a thing as an acute aortic dissection. Table 7.1 classifies the etiology, indications for surgery and technique of surgery for acute aortic dissection and chronic aortic aneurysm. Figure 7.1a-c provides a diagrammatic summary of the aortic pathologies and repair techniques.

AORTIC DISSECTION

Etiologies of acute aortic dissection include Marfan's syndrome, cystic medial necrosis, and hypertension. The presence of a bicuspid aortic valve has a high coincidence with aortic dissection. Acute aortic dissection arises from separation of the aorta within the media. The media and adventitia of the aorta literally dissect away from the intima and the inner portion of the media. Controlling the patient's hypertension is essential in slowing or halting the progression of the aortic dissection. There is always an intimal tear associated with the aortic dissection which permits blood to dissect into the media, and the location of this intimal tear is critical for planning surgical management. The tear is propagated simply by the pulsatile effect of cardiac contraction. Although the progression of the aortic dissection varies, usual locations are the right anterior portion of the ascending aorta, greater curvature of the transverse arch, and left anterolateral wall of the descending aorta.

Patients with acute aortic dissection typically present with a tearing sensation in the back which is different in quality and location than the pain of angina. Nonetheless, the pain of acute aortic dissection may mimic other disorders

---

Cardiothoracic Surgery, Third Edition, edited by Fritz J. Baumgartner. ©2004 Landes Bioscience.

Table 7.1.

	Etiology	Indications for Surgery	Technique
Acute Aortic Dissection	Marfan's (#1 etiology) Cystic Medial Necrosis HTN	Type A–mere presence	**Ascending**–50% have AI which must be managed by resuspension or AVR; can do Bentall, Cabrol, or separate AVR/ascending graft replacement.
	Bicuspid Aortic Valve	Type B–complications (rupture, visceral or limb ischemia, persistent pain)	**Descending**–Tube graft replacement; as with ascending type, can leave false lumen distally; may use LA-fem or fem-fem bypass to protect spine.
			Thoracoabdominal –Need LA-fem or fem-fem bypass to protect the spine.
			If aneurysm prox to innominate a., clamp and cannula may be applied to aorta.
Chronic Aortic Aneurysm	Atherosclerotic Disease (#1 Etiology) Marfan's Cystic Medial Necrosis	Ascending or Descending: > 6 cm Transverse or Thoracoabd	**Ascending** If distal to innominate a., RA-fem a. bypass then circ arrest. Use interposition or inclusion method; may need AVR by Bentall, Cabrol, or separate AVR/ascending graft.
	Aortitis (syphilis or granuloma)	Recent ↑ in size or symptoms	**Transverse:** RA-fem a. bypass then circ arrest. **Descending Thoracic:** 5th-7th interspace; LA → fem or fem-fem bypass. "clamp & go" if repair < 30 min **Thoracoabdominal:** 7-9th interspace; again use LA-fem a. bypass or fem-fem bypass.
Aortic Transection	Blunt Deceleration	Mere presence	If no prox extension, "clamp & go" if repair < 30 min otherwise LA-fem a. bypass. At ligamentum arteriosum (Left thoracotomy)
			If prox extension, LA → fem bypass, cool & circ arrest.
			Ascending Aorta (Median sternotomy) RA-fem a. bypass; clamp prox to innominate a. if possible but if not, need circ. arrest.

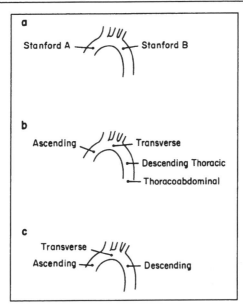

Fig. 7.1. Types of aortic pathologies and techniques of repair. a) Acute Dissection. Stanford A: RA-femoral a. bypass with circulatory arrest; If need to replace aortic valve, need Bentall/Cabrol or separate supraannular graft placement. Stanford B: surgery for medical failures; LA-femoral a. or fem-fem bypass to protect spinal cord. b) Chronic Aneurysm. Ascending: Standard CPB (cannulas in RA and aorta); Transverse: RA-femoral a. CPB with hypothermic circ. arrest (inclusion technique of graft placement; Descending Thoracic and Thoracoabdominal–LA-fem a. or fem. fem. bypass; circ. arrest is an alternative. c) Traumatic Rupture. Transverse RA–femoral a. bypass and circ. arrest; Ascending–Standard CPB; Descending–"clamp and go" or additional protection with LA-fem bypass; If tear in descending aorta goes proximally into arch, need fem-fem bypass and circ. arrest.

including angina, heartburn, or pneumothorax. The patient may present with the signs and symptoms of cardiac tamponade if rupture has occurred into the pericardium.

Figure 7.2 shows the types of aortic dissection based on the location. A dissection involving the ascending, transverse, and descending aorta is a DeBakey type I dissection; that involving only the ascending aorta is a DeBakey type II; a dissection only involving the descending thoracic aorta is a DeBakey type III dissection. This has been simplified in the Stanford classification, where one simply asks the question whether or not there is a component of an ascending aortic dissection. If the dissection does involve the ascending aorta, i.e., a DeBakey Class I or II, then this is a Stanford A dissection. If, however, only the descending thoracic aorta is involved (DeBakey III) then this is Stanford B. This is a simpler and clinically more useful classification because the management is exactly related to whether it is a type A or type B dissection. Type A dissections are managed with surgery and

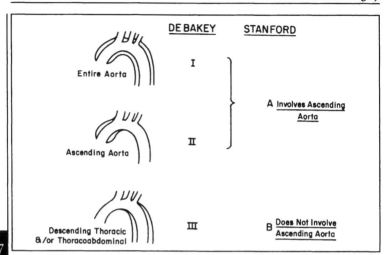

Fig. 7.2. Classification of aortic dissection.

type B dissections are generally managed only with antihypertensives and no surgery. The reason that type A dissections require surgery is because of the high risk of complications arising from the ascending aortic dissection. This includes dissection into the coronary ostia with infarction, rupture into the pericardium resulting in tamponade, acute aortic insufficiency from dissection into the aortic valve or free rupture and exsanguination. One year survival rates for patients with acute type A dissections are in the range of 60% for surgical management versus 5% for medical management; surgical management is clearly preferred. Conversely, medical management is superior for type B dissections, in the range of 70%, and is therefore the preferred treatment. Acute type A and B aortic dissections are depicted radiographically in Figs. 7.3 and 7.4, respectively.

Ascending aortic dissection is a dangerous condition that requires treatment. The ECHO, CT, and angiogram are useful in establishing the diagnosis of acute aortic dissection and evaluating its characteristics. ECHO delineates the extent and location of the intimal tear in the ascending aorta, including coronary artery involvement and aortic valve competence. In a patient with suspected acute aortic dissection, an ECHO alone will diagnose a type A dissection and mandate surgery. CT and angiograms, including coronary angiography, are generally not warranted preoperatively in an acute setting where the diagnosis is made by ECHO. The angiogram, however, remains the gold standard to delineate the extent and location of the intimal tear, location of the false lumen and specifics of the visceral vessels.

It is important to remember that on occasion there may be a type B dissection which progresses retrogradely to dissect up into the transverse and then ascending aorta even though the actual tear is in the aorta distal to the take-off of the left subclavian artery. This condition still requires surgery because it may further dis-

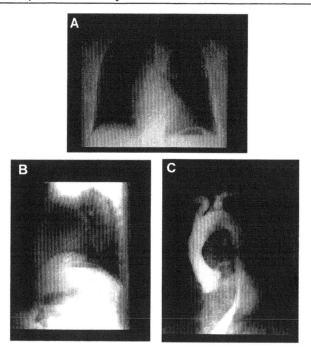

Fig. 7.3a-c. Acute type A dissection. PA (a) and lateral (b) chest films show an ectatic and dilated aorta. c) Aortogram of above patient reveals abnormal aortic narrowing throughout its length and aortic valve regurgitation.

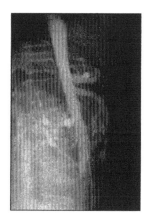

Fig. 7.3d. Visceral arteriogram reveals the celiac, mesenteric, and renal vessels arising from the true lumen. This patient underwent resuspension of the aortic valve and replacement of the ascending aorta. Courtesy Dr. Michael Janusz, University of British Columbia.

sect down to the coronaries or aortic valve or rupture into the pericardium. Blood pressure control is mandatory in these patients and includes not only vasodilator treatment with Nipride but also beta blockers to decrease the force of contraction,

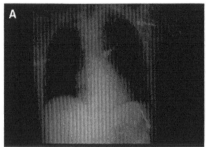

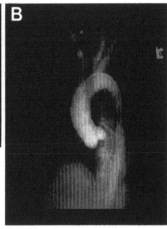

Fig. 7.4a. Acute type B dissection. Chest film (a) showing no specific indications of aortic dissection, although the aortic knob is somewhat blunted.

Fig. 7.4b. Aortogram showing narrowing of the true lumen of the aorta secondary to extrinsic narrowing by the false lumen.

Fig. 7.4c-d. CT scan at the level of the aortic arch and carina revealing a false lumen in the descending thoracic aorta. The proximal aorta is not involved.

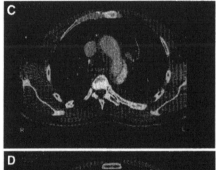

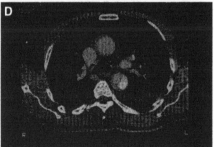

thus decreasing dP/dT and sheer force exerted on the dissection. Hypertensive therapy with beta blockers and vasodilators is the mainstay of management for

---

Table 7.2. Surgical decisions for type A aortic dissection

Management of Aortic Valve
- Resuspend valve
- AVR
- Leave aortic valve alone

Management of Coronary Arteries
- Reimplant coronaries
- Leave coronaries alone and place supracoronary graft

Management of Transverse Arch
- Place graft proximal to arch
- Replace arch

---

type B dissections. These patients generally do not require surgery unless their pain is not controlled after an appropriate course of antihypertensive management or if they have visceral sequelae of their dissection. This includes decreased blood flow to the renal arteries, celiac axis, superior mesenteric artery, or shearing off of the artery of Adamkiewitz with spinal ischemia or limb threat from shearing off of the iliac vessels.

Critical aspects in determining the surgical management of acute aortic dissection include whether or not there is coronary artery involvement, aortic insufficiency, and the presence of an intimal tear in the transverse arch (Table 7.2). Half of patients with type A dissections have aortic insufficiency which must be taken care of by resuspension (Fig. 7.5a-e) or aortic valve replacement. Aortic valve replacement for aortic dissection must be done as a Bentall or Cabrol procedure, or separate aortic valve and ascending graft replacements must be done.

A Bentall procedure (Fig. 7.6a-d) is a composite aortic valve replacement with replacement of the ascending aorta and reimplantation of the coronary artery orifices directly into the graft. The Cabrol is a simplified procedure in which the coronary orifices are not reimplanted directly into the graft, but rather a separate 8 mm Dacron graft is sewn onto both coronary orifices and then attached to the ascending aortic graft (Fig. 7.7a-b). This is technically easier than is separate reimplantation of the coronary orifices into the graft. Alternatively, a separate aortic valve replacement and ascending graft replacement can be done and the coronary orifices left alone without reimplanting them. If the coronary arteries are spared from the dissection, they need not be reimplanted into the graft. If they are involved, a Bentall or Cabrol procedure is required.

Where does one attach the graft distally? The dissection may continue around the aortic arch and even down into the descending thoracic aorta as with type I dissections. However, one does not need to go to the distal most extent of the dissection; one simply needs to eliminate the intimal tear. One can plicate the distal true and false lumens with felt and then attach this to the graft, thus obliterating the false lumen distally. For descending thoracic dissections, as for ascending, one can leave the dissection false lumen distally and simply plicate the true and false lumen distal to the intimal tear.

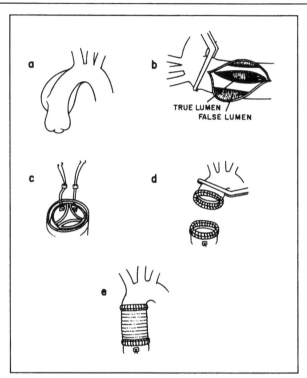

Fig. 7.5. Repair of type A aortic dissection with valve resuspension. a) Type A aortic dissection with aortic insufficiency. Right atrial-femoral artery bypass is established. b) Longitudinal aortomy revealing true and false lumens. c) Resuspension of aortic valve with two plegetted sutures at each commissure. d) Inner and outer felt secured at proximal and distal aorta using a whip stitch. e) Interposition tube graft placed between proximal and distal aorta.

TECHNIQUE FOR REPAIR OF ASCENDING AND TRANSVERSE ARCH DISSECTIONS

For acute aortic dissection involving the ascending aorta or transverse arch, right atrial-femoral artery bypass is established. During the process of actively cooling, a cross-clamp is applied to the aorta proximal to the innominate artery. The aortic root is opened and cardioplegia solution administered retrogradely or directly into the coronary orifices. At this point, a decision regarding the transverse arch must be made. If there is an obvious intimal tear in the ascending aorta and there is no real suspicion of a transverse arch intimal defect, then profound hypothermia and circulatory arrest is not required. Replacement of the ascending aorta and, if necessary, valve replacement or resuspension and coronary reimplantation can be performed. If, however, there is suspicion of transverse arch involvement, or if an intimal tear in the ascending aorta is not evident, or if the tear is too close to the aortic cross-clamp to permit an adequate sewing margin, then the patient must be cooled to 18°C in anticipation of circulatory arrest. While

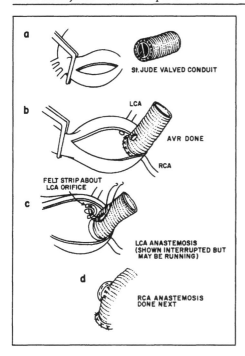

Fig. 7.6. Bentall procedure. a) Right atrial-femoral bypass and cardiac arrest are used. A longitudinal aortomy performed and the appropriate sized valved conduit chosen. b) The valve is seated. Coronary orifice holes are made in the tube graft with an ophthalmic electrocautery. c) Anastomosis of the left main coronary artery to the graft. Felt has been placed to reinforce the coronary ostium. d) Completed left main coronary anastomosis. The right coronary followed by distal anastomosis will be performed next.

7

cooling is underway, aortic valve resuspension or replacement is performed if necessary, and if coronaries must be reimplanted this is performed as well. With the patient cooled to 18°C, circulatory arrest is established. The patient is placed in Trendelenburg to prevent air from entering the cerebral vessels, and the cross-clamp is released. The arch is inspected and intimal tears are managed as direct suture plication, or reimplantation of the arch vessels as a separate pedicle or as part of a beveled distal anastomosis (Fig. 7.8). De-airing is performed by turning on the pump slowly, de-airing the aorta from the transverse arch and ensuring air does not enter the arch vessels. Full bypass is re-established and the patient re-

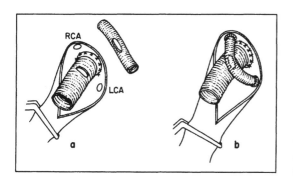

Fig. 7.7. Cabrol procedure. a) The aortic valve has been excised and the valved conduit seated in position. An 8 mm Dacron graft is used to anastomose the right and left coronary orifice to the valved conduit. b) The distal aortic anastomosis is performed next.

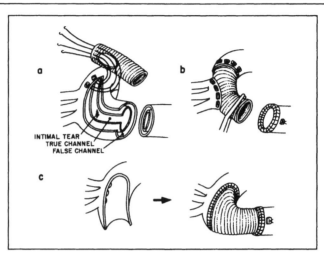

Fig. 7.8. Repair of arch dissection. a) Bypass, cardiac arrest, and profound hypothermia with circulatory arrest are established. A graft is inserted with separate anastomosis of the distal aorta and arch vessels. The intimal tear is excluded. b) The cross-clamp is applied to the proximal graft, and cardiopulmonary bypass and warming are resumed. The arch vessels are thus perfused while the proximal anastomosis is performed. c) Alternatively, a beveled anastomosis is performed to include both the distal aorta and arch vessels.

warmed while any unfinished proximal procedures (e.g., aortic valve replacement, coronary reimplantation) are performed.

Several technical points should be made. Generally for Marfan's, an aortic valve replacement as a valve conduit (Bentall or Cabrol procedure) should be done rather than a supracoronary conduit and separate aortic valve replacement. This is because of the chance of degeneration and aneurysmal dilation of the aortic segment containing the coronary orifices. If the dissection is not due to Marfan's syndrome, then an attempt is made to resuspend the valve and place a supracoronary graft without reimplanting the coronary arteries. The texture and "feel" of the aorta is an important factor in the surgeon's decision making process for the use of felt. Acute aortic dissections are extremely friable compared to the sturdier aneurysm walls and require meticulous suturing and felt reinforcement. For aortic dissection, a felt strip is placed within the aorta as well as another strip outside the aorta and secured with a whip stitch, then the Dacron tube graft separately sewn onto this (Fig. 7.9a). Another technique is to incorporate both the graft, outer felt, aorta, and inner felt as one continuous whip stitch (Fig. 7.9b).

Another technical point is that when performing the distal anastomosis in aortic dissection, the pump should be slowly turned on to be sure one is performing the anastomosis to the true lumen. If the true lumen has been sewn shut, no blood will be seen to come out of the lumen. In this case, a V-excision must be made between the true and false lumens as a type of fenestration.

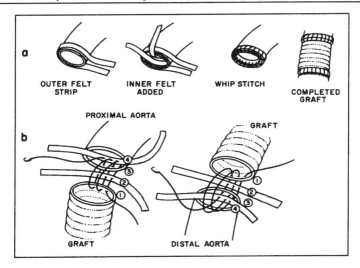

Fig. 7.9. Methods of felt repair in acute aortic dissection. a) Separate preparation of aortic ends with felt. An outer and inner layer of felt are anchored to the aorta with a whip stitch, and the graft is anastomosed to this. b) Combined graft anastomosis with addition of felt. The entire anastomosis may be done at once with both strips of felt. The suture is placed from graft to outer felt to aorta to inner felt. 1: graft; 2: outer felt; 3: aorta; 4: inner felt.

Cerebral ischemia during the period of circulatory arrest is of paramount concern. Circulatory arrest times greater than 45-60 minutes dramatically increase the chances of neurologic sequelae. Techniques of cerebral protection include systemic cooling to 18°C, packing the head in ice, and intraoperative administration of mannitol, steroids, and free radical scavengers. Cerebral perfusion may be performed, either by the antegrade or retrograde route. In the antegrade route, a great vessel or one of its branches (e.g., right subclavian artery) is cannulated and cold blood administered during the circulatory arrest period. In the retrograde approach, the superior vena cava is cannulated and snared and cold blood administered retrogradely. These techniques appear to increase the "safe" period of circulatory arrest.

The tissues in acute aortic dissection are extraordinarily fragile, and felt strips have been historically used most often in the United States to reinforce these tissues. Several surgical glues are used in acute aortic dissections and have a threefold purpose: reinforcement of friable tissues, support of suture lines for hemostasis and obliteration of the false lumen in aortic dissection. Gelatin-resorcinal-formaldehyde (GRF) is the glue that has previously been used most commonly. Although the glue has a good track record, there are concerns about the high reoperation rate and tissue toxicity. An alternative is the recently introduced Bioglue Surgical Adhesive (CryoLife International, Inc., Kennesaw, GA, U.S.A.) consisting of 10% glutaraldehyde and 45% bovine serum albumin. Mixing the components occurs in situ at surgery, creating a strong scaffold through immediate polymer-

Fig. 7.10. Relative incidence of thoracic aneurysms.

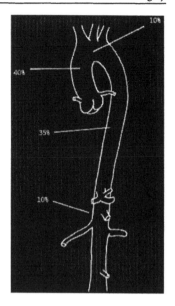

ization and binding of lysine molecules. Within 2 minutes maximal binding support is achieved. Most surgeons feel these surgical glues greatly facilitate surgery for acute aortic dissection by strengthening tenuous tissues, obliterating the false lumen and improving hemostasis.

THORACIC AORTIC ANEURYSMS

GENERAL CLASSIFICATION AND PRINCIPLES

Thoracic aneurysms are most commonly located in the ascending and descending thoracic regions (Fig. 7.10). The more difficult transverse arch and thoracoabdominal locations are also less common. Several etiologies exist for chronic aortic aneurysms. The most common cause is atherosclerotic degeneration of the aorta (Fig. 7.11a-d), usually occurring in elderly smokers. Chronic aortic dissection (Fig. 7.12a-c), Marfan's syndrome (Fig. 7.13a-c), cystic medial necrosis or aortitis secondary to syphilis or granuloma are other etiologies.

The indications for surgical correction include a diameter greater than 6 cm, or a dramatic increase in size or symptoms over a short time interval. Those generally portend a higher chance of rupture and poor prognosis without surgery. Smaller aneurysms can generally be followed, although a size between 5-6 cm represents a gray zone. The presence of aortic insufficiency is a compelling reason for aortic replacement even for smaller aneurysms.

If an aneurysm of 5 cm diameter is found incidentally in the ascending aorta while performing another cardiac procedure, one of several acceptable choices

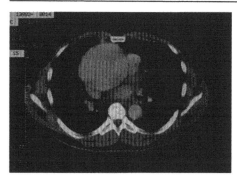

Fig. 7.11a. CT scan of a patient with a 9 cm aneurysm confined to the ascending aorta resulting in aortic insufficiency.

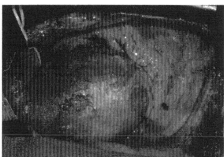

Fig. 7.11b. Aneurysm as seen at sternotomy. Notice the normal sized distal ascending aorta encircled by vascular tape. Cannulation of the normal aorta and cardiopulmonary bypass are performed. Courtesy of Dr. Jeff Milliken, Harbor-UCLA

Fig. 7.11c. The aorta has been opened, aortic valve excised, and annular sutures placed. A St. Jude valved conduit will be positioned and the Cabrol procedure performed. Courtesy of Dr. Jeff Milliken, Harbor-UCLA

Fig. 7.11d. The valved conduit has been seated in the aortic annulus. An 8 mm graft has been anastomosed to the left and right coronary orifices and now will be sutured to the valved conduit. Courtesy of Dr. Jeff Milliken, Harbor-UCLA

Fig. 7.12a. CT scan of a patient with a chronic type B dissection resulting in an aortic aneurysm. The intimal flap is clearly seen in both the ascending and descending thoracic aorta. The patient had severe aortic insufficiency and coronary artery disease as well.

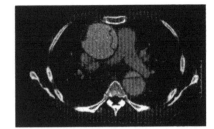

Fig. 7.12b. The chronic dissection continues into the abdominal aorta, here seen at the level of the superior mesenteric artery.

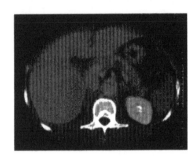

Fig. 7.12c. The right common iliac artery has been obliterated by the chronic false lumen. The patient underwent repair of the ascending and transverse aneurysm, with circulatory arrest, resuspension of the aortic valve and coronary bypass.

must be made. There is justification for either not performing any aortic intervention or alternatively placing Dacron mesh around the aorta for additional support. Some advocate replacement of the ascending aorta at this point, although this is usually not warranted. A reduction aortoplasty is also a valid alternative.

TECHNIQUE OF REPAIR

For ascending aortic aneurysms, if the aneurysm does not extend beyond the origin of the innominate artery, one may be able to place the arterial cannula in the distal ascending aorta or transverse arch and initiate cardiopulmonary bypass

Fig. 7.13a. The aneurysmal ascending aorta of a 3 year old child with Marfan's syndrome. The cross-clamp can still be applied to the distal ascending aorta and bypass achieved into the transverse arch. Courtesy of Dr. Jeff Milliken, Harbor-UCLA

Fig. 7.13b. The aorta is opened and the regurgitant aortic valve excised. Courtesy of Dr. Jeff Milliken, Harbor-UCLA.

7

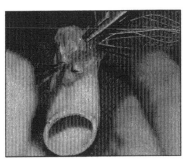

Fig. 7.13c. A homograft is prepared for aortic root replacement. Annular sutures are being placed into the homograft. Courtesy Dr. Jeff Milliken, Harbor-UCLA.

with inflow from the right atrium. A cross-clamp is placed proximal to the innominate artery, and adequate perfusion of the arch vessels is achieved (Fig. 7.11a-d, 7.13a-c). If the ascending aortic aneurysm extends distal to the innominate, it will be impossible to properly clamp proximal to the innominate artery. In this case, right atrial-femoral arterial cardiopulmonary bypass is established.

As for acute aortic dissection, several well-defined management decisions must be made with regard to the aortic valve, coronary arteries, and transverse arch. If the aortic valve must be repaired or replaced but the coronaries are spared, a separate supracoronary tube graft may be placed. As mentioned in the section on acute dissection, for Marfan patients, a valve replacement should be done as a valved conduit rather than as an isolated aortic valve replacement and supracoronary conduit. This is to prevent the chance of degeneration of the remaining aortic segment with the coronaries. If the coronaries must be reimplanted, a Bentall or Cabrol procedure is performed (Fig. 7.11a-d, 7.13a-c). If the transverse arch is

Fig. 7.14a. Transverse arch aneurysm repair. a) Interposition technique with great vessels attached to graft as a separate patch or included within the distal anastomosis.

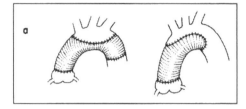

Fig. 7.14b. Inclusion technique sewing within the opened aorta.

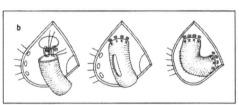

Fig. 7.14c. CT scan of localized transverse arch aneurysm to undergo simple patch repair.

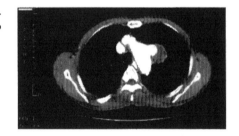

involved, it is replaced under profound hypothermia and circulatory arrest (Fig. 7.14a,b). There are some instances of localized transverse arch aneurysms which can be managed by patch repair without circulatory arrest (Fig. 7.14c).

For descending thoracic aneurysms, the level of the thoracotomy depends on the level of the aneurysm. Generally, a left fifth interspace thoracotomy is performed. Partial bypass using a centrifugal pump is used to route oxygenated blood from the left atrium to the femoral artery (Fig. 7.15). This enables perfusion of the spinal artery of Adamkiewitz to limit the incidence of paraplegia. Only partial heparinization is required (1 mg/kg). The mortality and paraplegia incidence for the repair of a descending thoracic aneurysm is on the order of 10% and 5% respectively. More recently profound hypothermia with circulatory arrest has been used for descending thoracic aneurysms with great success. Another alternative for managing aortic aneurysms is endocascular stent placement. These generally work best for localized descending thoracic aneurysms and are a logical extension of the extensive experience with infrarenal abdominal aortic aneurysms.

Thoracoabdominal aneurysms are significantly more dangerous, with repair resulting in published mortalities of 10-50% and incidence of paraplegia 7-25%. The location of these aneurysms and their proximity to and involvement of the spinal artery of Adamkiewicz and visceral vessels make them particularly treacherous (Fig. 7.16). The repair of such an aneurysm requires a thoracoabdominal

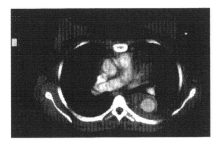

Fig. 7.15a. CT scan of a 35 year old female with a large descending thoracic aneurysm with mural clot. Courtesy of Dr. Jeff Milliken, Harbor-UCLA.

Fig. 7.15b. A left thoracotomy reveals the large-sized aneurysm. Courtesy of Dr. Jeff Milliken, Harbor-UCLA.

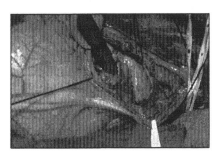

Fig. 7.15c. Left atrial-femoral partial bypass is established. The left atrial cannula is seen in position. The vagus and phrenic nerves are retracted with vascular tapes. Courtesy of Dr. Jeff Milliken, Harbor-UCLA.

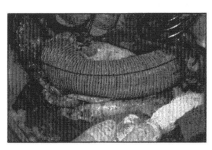

7.15d. Completed aneurysm repair. Courtesy of Dr. Jeff Milliken, Harbor-UCLA

7

incision. The patient is placed in the lateral decubitus position with the pelvis corkscrewed to gain access to the abdomen and groins. A low thoracic approach (usually the ninth interspace) is utilized, although if the aneurysm arises proximally near the left subclavian, it may be necessary to perform a separate thorac-

Fig. 7.16a-b. Chest and abdominal CT of a large thoracoabdominal aneurysm. The aneurysm was repaired under left atrial-femoral bypass. The proximal anastomosis was done through a left fourth interspace approach. The distal anastomosis was done through a left ninth thoracoabdominal approach.

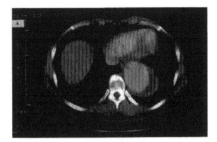

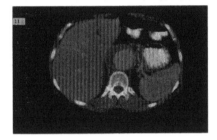

7

otomy in the fourth interspace to achieve exposure for the proximal anastomosis. The thoracoabdominal incision is extended to the midline of the abdomen and the diaphragm is entered along its periphery.

A retroperitoneal approach is used and the spleen, kidney and left colon are mobilized anteriorly and to the right to expose the aorta. Left atrial to femoral artery bypass is established. The aorta is cross-clamped proximal to the aneurysm and distally at a level above the visceral vessels (Figs. 7.17a-e). This permits continued perfusion of visceral and spinal vessels via the left atrial to femoral bypass circuit and decreases the chance of paraplegia. When the proximal anastomosis has been performed, the proximal clamp is taken off with the graft clamped to check the proximal anastomosis. The distal clamp above the visceral vessels is then removed after the left atrial femoral bypass circuit has been shut off. The visceral vessels are then reimplanted, usually with a celiac, superior mesenteric and right renal artery pedicle, and with the left renal as another pedicle. The cross-clamp is placed on the distal portion of the graft and the proximal clamp on the aorta is removed. This allows perfusion of the visceral and spinal vessels as the distal most anastomosis is performed. Alternatively, after the proximal anastomosis has been performed, the distal anastomosis is performed as one large spatulated anastomosis to include all the visceral vessels (Fig. 7.18). Not all thoracoabdominal aneurysms are anatomically suited to this simpler repair. Again, profound hypothermia and circulatory arrest have been advocated as alternatives to left atrial-femoral bypass again with excellent results.

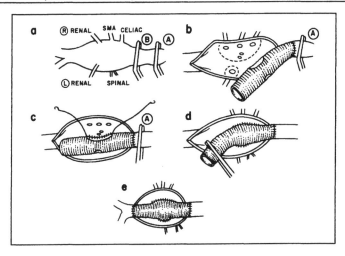

Fig. 7.17. Thoracoabdominal aneurysm resection. a) Left atrial-femoral bypass established. Cross-clamps placed proximally at (A) and (B) while perfusing visceral and spinal vessels. The proximal anastomosis is performed during left atrial-femoral bypass. b) After the proximal anastomosis has been performed, the bypass pump is turned off, the clamp at (B) is removed and the clamp at (A) is removed and placed on the graft. c) The visceral vessels are implanted in the graft. d) After the visceral vessels are reimplanted, the distal graft is clamped and the proximal clamp at (A) is removed. The distal anastomosis is performed while the visceral and spinal vessels are reperfusing. e) Completed repair.

Surgical judgment and modification of techniques tailored to meet individual needs is a necessity. For example, separate reimplantation of a button of spinal arteries may be necessary. In cases of "mega-aorta syndrome" (Fig. 7.19a-e), the entire aorta including ascending, transverse, descending thoracic, and

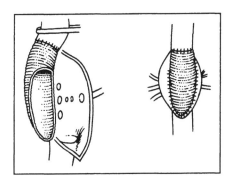

Fig. 7.18. Thoracoabdominal aneurysm resection: inclusion of visceral vessels within the distal anastomosis.

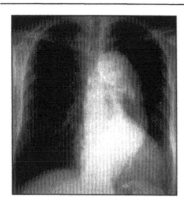
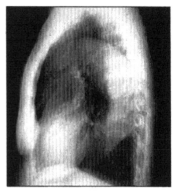

Fig. 7.19a-b. Chest films of a patient with mega-aorta syndrome and aortic insufficiency showing the massively dilated aorta.

Fig. 7.19c. The aneurysmal portion includes the ascending and transverse aorta.

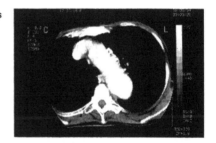

Fig. 7.19d. Descending aorta.

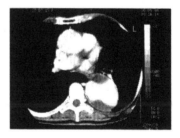

Fig. 7.19e. Thoracoabdominal aorta. Anatomically, this is an ideal candidate for a staged elephant trunk procedure. Courtesy Dr. Hans Borst, Medizinische Hochschüle, Hannover.

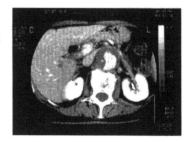

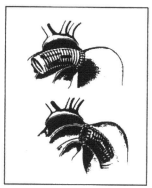

Fig. 7.20a. Elephant trunk procedure. a) The graft is inverted upon itself retrogradely into the arch graft. The distal elephant trunk portion is then pushed back down the descending thoracic aorta. Alternatively, the future arch graft is invaginated distally into the trunk and then extended out into the arch.

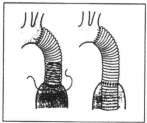

Fig. 7.20b. For distal elephant trunk insertion, the future trunk is inverted retrogradely into the descending aortic graft, then pulled downstream just prior to completing the distal anastomosis.

7

thoracoabdominal aorta may be aneurysmal. The "elephant trunk" procedure originally conceived by Hans Borst utilizes a staged sequence of aortic operations with each subsequent step simplified by the presence of a free-floating distal end of graft (Fig. 7.20a-b). On subsequent redo operations, achieving control of the distal graft is a simple matter (as opposed to taking down a distal graft to aorta anastomosis).

TRAUMATIC AORTIC RUPTURE

Aortic transection or other forms of traumatic aortic injury usually occur from a deceleration type of blunt trauma and usually occurs when a patient smashes his chest against the steering wheel in a motor vehicle accident. The tear usually occurs at the region of the ligamentum arteriosum just distal to the left subclavian artery. In this case, a left thoracotomy is performed and if there is a small hematoma without proximal extension, one can clamp the aorta proximal to the left subclavian, separately clamp the left subclavian, clamp the distal aorta and then do the repair with interposition graft if the cross-clamp time can be kept to less than 30 minutes. Cross-clamp time greater than 30 minutes results in an increased chance of paraplegia. If there is a large hematoma with proximal extension into the transverse or ascending aorta, femoral-femoral bypass is established, the patient is cooled, circulatory arrest established, the aorta opened and the repair done. This is achieved via a left thoracotomy. For ascending aortic traumatic injuries, a median sternotomy is performed and right atrial-femoral bypass is established. If one can get a cross-clamp proximal to the innominate, then one can perform the procedure with the aortic cannula in the distal ascending aorta or in the transverse arch. If one cannot get a cross-clamp on proximal to the innominate, then right atrial-femoral artery bypass is established, the patient cooled and hypothermic circulatory arrest established to perform the repair. Aortic transection is further discussed in Chapter 18 "Thoracic Trauma".

SUGGESTED READING

1. Crawford ES, Crawford JL. Diseases of the aorta including an atlas of angiographic pathology and surgical technique. Baltimore: Williams & Wilkins, 1984.
2. Cabrol C, Pavic A, Mesnildrey P et al. Long-term results with total replacement of the ascending aorta and reimplantation of the coronary arteries. J Thorac Cardiovasc Surg 1986; 91:17-25.
3. Kouchoukos NT, Marshall WG Jr, Wedge-Stecher TA. Eleven-year experience with composite graft replacement of the ascending aorta and aortic valve. J Thorac Cardiovasc Surg 1986; 92:691-705.
4. Crawford ES, Saleh SA. Transverse aortic arch aneurysm: Improved results of treatment employing new modifications of aortic reconstruction and hypothermic cerebral circulatory arrest. Ann Surg 1981; 194:180-188.
5. Crawford ES, Snyder DM. Treatment of aneurysms of the aortic arch. J Thorac Cardiovasc Surg 1983; 85:237-246.
6. Griepp RB, Stinson EB, Hollingsworth JF et al. Prosthetic replacement of the aortic arch. J Thorac Cardiovasc Surg 1975; 70:1051-1063.
7. DeBakey ME, McCollum CH, Crawford ES et al. Dissection and dissecting aneurysms of the aorta: twenty year follow-up of five hundred twenty-seven patients treated surgically. Surgery 1982; 92:1118-1134.
8. Crawford ES, Crawford JL, Safi HJ et al. Thoracoabdominal aortic aneurysms: Preoperative and intraoperative factors determining immediate and long-term results of operations in 605 patients. J Vasc Surg 1986; 3:389-404.
9. Svensson LG, Crawford ES, Hess KR et al. Variables predictive of outcome in 832 patients undergoing Repairs of the descending thoracic aorta. Chest 1993; 104:1248-53.
10. Borst HG, Jurmann MJ, Bukner B et al. Risk of replacement of descending aorta with a standardized left heart bypass technique. J Thorac Cardiovasc Surg 1994, 107:126-33.
11. Lawrie GM, Erale N, DeBakey ME. Long-term fate of aortic root and valve after ascending aneurysm surgery. Am Surg 1993; 217:711
12. Lass J, Jurmann MJ, Heinemann M et al. Advances in aortic arch surgery. 1992; 53:227-32
13. Crawford ES, Kirklin JW, Naftel DC et al. Surgery for acute ascending aortic dissection: Should the arch be included? J Thorac Cardiovasc Surg 1992; 104:46-59.
14. Coselli JS, Büket S, Djakanovic B. Aortci arch surgery: Current treatment and results. Ann Thorac Surg 1995; 59:19-27.
15. Najafi H. 1993 update: Descending aortic aneurysmectomy without adjuncts to avoid ischemia. Ann Thorac Surg 1993; 55:1042-5.

16. Yun KL, Glower DD, Miller DC et al. Aortic dissection resulting from tear of transverse arch: Is concomitant arch repair warranted? J Thorac Cardiovasc Surg 1991; 102:355.

17. Svensson LG, Crawford ES. Aortic dissection and aortic aneurysm surgery: Clinical observations, experimental investigations, and statistical analyses. Part II. Curr Probl Surg 1992; 29:913.

18. Kato M, Ohnishi K, Kaneko M et al. New graft-implanting method for thoracic aortic aneurysm or dissection with a stented graft. Circulation 1996; 94(Suppl2):188-93.

19. Dake MD, Miller DC, Semba CP et al. Transluminal placement of endovascular stent grafts for the treatment of decending thoracic aortic aneurysms. N Engl J Med 1994; 331:1729-34.

20. Fann JI, Dake MD, Semba CP et al. Endovascular stent-grafting after arch aneurysm repair using the "elephant trunk". Ann Thorac Surg 1995; 60:1102-5.

21. Yano H, Ishimaru S, Kawaguchi S. Endovascular stent-grafting of the descending thoracic aorta after arch repair in acute type A dissection. Ann Thorac Surg 2002; 73:288-91.

22. Passage J, Jalali H, Tam RKW et al. Bioglue surgical adhesive—an appraisal of its indications in cardiac surgery. Ann Thorac Surg 2002; 74:432-7.

7

Hypertrophic Obstructive Cardiomyopathy

Bassam O. Omari, Fritz J. Baumgartner

HYPERTROPHIC OBSTRUCTIVE CARDIOMYOPATHY

Hypertrophic obstructive cardiomyopathy, also known as idiopathic hypertrophic subaortic stenosis (IHSS), is a autosomal dominant disease in which there is hypertrophy of the interventricular septum classically in the cephalad portion, hence the term asymmetric septal hypertrophy. Systolic anterior motion of the anterior leaflet of the mitral valve occurs due to the free edge of the anterior leaflet bulging close to the septum, causing an abnormal motion seen on echocardiography. An echocardiogram is the most helpful diagnostic maneuver in evaluating IHSS. If the patient is refractory to medical treatment (Inderal, calcium channel blockers and diuretics), then the patient should undergo cardiac catheterization with a view to surgery.

The treatment of hypertrophic obstructive cardiomyopathy is transaortic myomectomy. If the extent of intraventricular hypertrophy is deeper in the left ventricular outflow tract than usual, or when residual muscle is left behind, a left ventricular approach is used also. Cardiopulmonary bypass is established with a single venous cannula, a left ventricular vent is used and antegrade followed by retrograde cardioplegia is used to arrest the heart using cold blood cardioplegia. The aortic root is opened. A sponge is placed over the right ventricle to push the septum down and an incision is made down the interventricular septum from a point just to the left of the nadir of the right coronary cusp and continued toward the intercoronary commissure in the "Morrow procedure". A wedge of muscle is thus excised.

Generally, surgical treatment is reserved for patients in Class III or IV failure, angina or syncope, or those with a gradient greater than 50 mm, i.e., similar to that for aortic stenosis, or those patients not relieved with Inderal, calcium channel blockers or diuretics. Even if there is no gradient at rest, if there is a greater than 50 mm gradient on exercise, surgery is indicated.

Mitral valve replacement may be necessary for some cases of IHSS but this is only done when myomectomy has not decreased the dynamic left ventricular outflow tract gradient. Left bundle branch block occurs in up to two-thirds of patients, but a complete AV block should not occur because one stays to the left of the nadir of the right coronary cusp and, hence, away from the membranous septum and the bundle of His at the anterior commissure.

Cardiothoracic Surgery, Third Edition, edited by Fritz J. Baumgartner. ©2004 Landes Bioscience.

CARDIAC TUMORS

Myxomas are the most common benign tumors of the heart. They are usually globular in shape and have a soft, gelatinous consistency. They have a predilection for the left atrium where they are attached to the limbus of the fossa ovalis.

Myxomas occur in patients of all age groups but are most frequent in middle-aged women. Seventy-five percent of myxomas originate in the left atrium, 18% in the right atrium and the remainder in the ventricles.

Patients present either with signs and symptoms of hemodynamic compromise, with systemic embolization, or rarely with constitutional symptoms. The hemodynamic symptoms are usually due to obstruction of flow across the atrioventricular valves. Patients with left atrial myxomas often present with dyspnea, orthopnea, and paroxysmal nocturnal dyspnea, with or without hemoptysis, as in mitral stenosis. This obstruction of flow is characteristically progressive but could be intermittent. Myxomas may cause atrioventricular valve insufficiency by preventing valve closure or by damaging the valve leaflets.

Systemic embolization is a major complication of left atrial myxomas and this is the presentation of almost 50% of the patients. Emboli have been reported in every organ including the coronary arteries. About 50% of emboli involve the central nervous system resulting in various neurological deficits. Peripheral arterial emboli result in pain and coldness of an extremity. Coronary artery embolization causes angina. Pulmonary embolism from right-sided myxomas is surprisingly low.

The constitutional symptoms from myxomas include fever, malaise, weight loss, clubbing of the fingers and toes, fatigue, myalgias and arthralgias. These constitutional symptoms are not infrequently the only manifestations of cardiac myxomas.

Two-dimensional echocardiography is now the most appropriate screening and diagnostic imaging modality for cardiac myxomas (Fig. 8.1). This method can identify the precise origin of the myxoma and it may demonstrate the tumor prolapse through the atrioventricular valve (Fig. 8.2a-b). It also can exclude any valvular heart disease.

Surgical excision is the treatment of choice of cardiac myxomas and this should proceed on an urgent basis. This excision is done under direct vision using cardiopulmonary bypass and hypothermic cardioplegic arrest. Tumor manipulation should be minimal during cannulation to prevent tumor dislodgement and embolization. The operative mortality is less than 5%, and the prognosis after surgical excision is usually excellent.

Although the most common benign primary cardiac tumor by far is the myxoma, other benign tumors include rhabdomyoma, fibroma, lipoma, teratoma, hemangioma, and pheochromocytoma. The papillary fibroelastoma is a small, benign neoplasm usually on the midportion of the aortic or mitral valves, but may occur on the papillary muscles, chordae, or elsewhere.

Cardiac malignancies are primary or metastatic, and while both are very rare, the latter are exceedingly more common. Of the infrequent primary cardiac

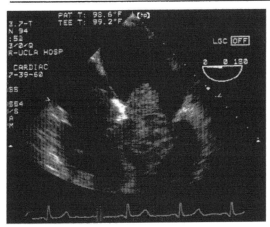

Fig. 8.1. A transesophageal echocardiogram (TEE) showing a left atrial myxoma attached to the atrial septum. Courtesy of Leonard Ginzton, Harbor-UCLA, UCLA School of Medicine.

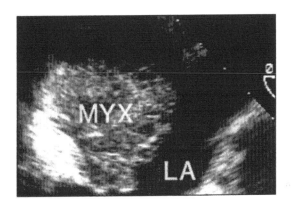

Fig. 8.2. A TEE showing left atrial myxoma in systole (a) and diastole (b). Note prolapsing of this myxoma through the mitral valve in diastole. Courtesy of Leonard Ginzton, Harbor-UCLA, UCLA School of Medicine.

Fig. 8.2a.

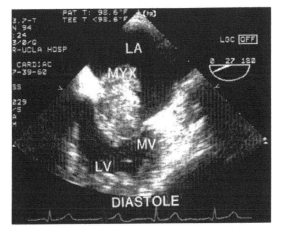

Fig. 8.2b.

malignancies, angiosarcoma is most frequent, usually as a bulky mural mass in the right atrium with frequent metastases. Rhabdomyosarcoma is the next most common and does not have a propensity for one cardiac chamber over another. Rarer malignancies include mesothelioma, plasmocytoma, and osteo-, lipo-, fibro-, and leiomyosarcoma.

SUGGESTED READING

1. Chitwood WR Jr. Cardiac neoplasms: Current diagnosis, pathology and therapy. J Cardiac Surg 1988; 3:119.
2. Dein JR, Frist WN, Stinson EB et al. Primary cardiac neplasms. J Thorac Cardiovasc Surg 1987; 93:502.
3. Semb BKH. Surgical considerations in the treatment of cardiac myxoma. J Thorac Cardiovasc Surg 1984; 87:251.
4. Pechacek LW, Gonzalez-Camid F, Hall RJ et al. The echocardiographic spectrum of atrial myxoma: A ten-year experience. Tex Heart Inst J 1986; 13:179.
5. Morrow AG. Hypertrophic subaortic stenosis. Operative methods utilized to relieve left ventricular outflow obstruction. J Thorac Cardiovasc Surg 1978; 76:423-430.
6. Maron BJ, Barrow RO, Cannon RO et al. Hypertrophic cardiomyopathy. Interrelation of clinical manifestations, pathophysiology, and therapy. N Engl J Med 1987; 316:844-852.
7. McIntosh CL, Greenberg GH, Maron BJ et al. Clinical and hemodynamic results after Mitral valve replacement in patients with obstructive hypertrophic cardiomyopathy. Ann Thorac Surg 1989; 47:236-246.

Minimally Invasive Heart Surgery

Fritz J. Baumgartner

The advent of laparoscopic cholecystectomy in the late 1980s conceptually changed surgeons' perceptions of operative procedures. The impact of these new perceptions affected not only general surgery, but most surgical specialties, and thoracoscopy was a natural extension of laparoscopic procedures. Cardiac surgery, however, was considered by most to be insulated from a general acceptance of these new "minimally invasive" procedures. This concept has radically changed in the last several years.

A primary question regards the very definition of "minimally invasive" with respect to cardiac surgery. Extrapolated from laparoscopic surgery, "minimally invasive" cardiac surgery implied a small incision. However, limited exposure incisions in cardiac surgery can lead to dire consequences if a misadventure occurs. The end result of a small, cosmetic incision does not justify turning a routine cardiac procedure into a risk to the patient's well being, or inordinately prolonging the length of time to complete the procedure.

Furthermore, endoscopic gallbladder surgery compared to open cholecystectomy is a physiologically poor analogy when applied to cardiac surgery because the heart-lung machine adds a new level of physiologic derangement. Cardiopulmonary bypass (CPB) and ischemic arrest are nonphysiologic as attested to by systemic inflammation, coagulopathy, low vascular resistance, and low-output postpump states. Furthermore, neuropsychiatric events and outright stroke are known risk factors of CPB which increase markedly with age. Adverse cerebral outcomes from CPB occur in about 6% of all patients undergoing coronary artery bypass (CABG), with 3% being type I events (strokes, transient ischemic attacks) and 3% being type II events (deterioration of intellectual function or seizures). However, when the specific subgroup of patients greater than 80 years of age undergoing CABG with CPB is examined, the adverse cerebral complication rate increases to about 16%, with about 8% each for type I and type II events. There are therefore compelling reasons to limit CPB time in the elderly population.

Besides the elderly, other subgroups who have a higher incidence of adverse outcomes from CPB include patients with profound ventricular dysfunction, prior stroke, and pulmonary and renal dysfunction. In patients such as these, a smaller incision may have less importance in the definition of "minimally invasive" than does limiting the cross-clamp and pump times. If a smaller incision means that these times are prolonged, then these "minimal incision" procedures should not necessarily be deemed as "minimally invasive."

It is for these reasons that another school of thought maintains that "minimally invasive" for heart surgery should be viewed in terms of total-body trauma rather than local in terms of incision size. These surgeons generally maintain that the key element in "minimally invasive" heart surgery is reducing, or eliminating, the use of CPB rather than using small incisions. For coronary bypass surgery, this means performing grafts on coronary vessels while the heart is beating. Valve surgery, which by its very nature requires entry into the cardiac chambers, requires CPB, and some would therefore consider "minimally invasive" valve surgery to be a misnomer. Some of the more common "minimally invasive" cardiac procedures follow.

"MINIMALLY INVASIVE" VALVE SURGERY

The term "minimally invasive" for valve surgery refers to incision size and approach, rather than limiting cross-clamp and perfusion times, since cardiopulmonary bypass is necessary. Several years ago alternative approaches to a standard median sternotomy were introduced as an improvement based on presumed decreased pain and shortened hospital stays. The initial approach introduced was a right parasternal approach to gain access to the aortic, mitral and tricuspid valves. Pain, because of the thoracotomy and excision of costal cartilages, as well as limitation of exposure if a problem arose limits the usefulness of this parasternal approach. The procedure evolved to a sternal transection to approach the aortic valve with transection of both internal mammary arteries. The reported justification for transection of both internal mammary arteries was that patients requiring aortic valve replacement alone rarely presented later for coronary revascularization. In terms of sternal devascularization, it was felt that retrograde and collateral flow via the distal internal mammary would be sufficient to not increase the risk of infection. This further evolved into a partial upper median sternotomy for aortic, mitral and tricuspid valves, with hemi-transection of the sternum into the right 3rd or 4th interspace, or a "J" type of sternotomy (Fig. 9.1). An alternative approach is the "J" incision made from the right 1st interspace to the midline of the sternum, and then down to the right 4th or 5th interspace.

These partial sternotomy incisions are superior to the right thoracotomy because of decreased pain and the ease with which the partial sternotomy can be converted to a full sternotomy should the need arise. It remains unresolved whether indeed these minimal incisions truly decrease patient morbidity or hospital stays. Cardiopulmonary bypass times are generally at least as long as their full sternotomy counterparts.

Another approach adopted several years ago involves Port-Access technology, wherein a very small thoracic incision is constructed to perform valve (or coronary) surgery. Fully endoscopic heart surgery has been possible with the technique. CPB is achieved via the femoral artery and vein route. A balloon catheter retrogradely placed from the femoral artery and inflated in the ascending aorta serves as a functional cross-clamp and antegrade cardioplegia delivery device. Coro-

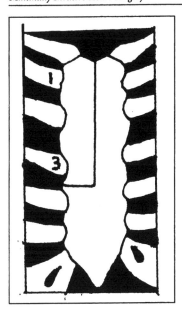

Fig. 9.1. Minimal "J" sternal incision for exposure of cardiac valves.

nary sinus cannulation for retrograde cardioplegia is endovascularly achieved via a central vein under transesophageal echocardiographic guidance. Port-Access has made possible the feat of completely endoscopic valve (and coronary) surgeries. Again, the benefit remains primarily incision size, since the bypass and ischemic times generally exceed their more standard counterparts. Furthermore, another incision is required in the groin to access the femoral vessels. The complexity of the various cannulations and need for fluoroscopy limit wide applicability as well.

MIDCAB

Minimally invasive direct coronary artery bypass (MIDCAB) was introduced several years ago as truly minimally invasive cardiac surgery, in terms of not only incision size, but, far more importantly physiologically, the fact that CPB was not required. The key to the reliability of the procedure is a stabilizing device shaped like a sewing machine foot pad which straddles the target artery (Fig. 9.2). This immobilizes the region, permitting an anastomosis on a relatively fixed target. MIDCAB generally refers to left anterior descending artery (LAD) grafting with the left internal mammary artery (LIMA) via small left anterior thoracotomy. Prior to development of a reliable stabilizing device, anastomoses to vessels on a vigorously beating target required pharmacologic manipulation to slow, or even transiently arrest the heart rate. Even so, these beating heart cases led in some instances to unsatisfying anastomoses. The use of a reliable stabilizing device was a seminal event in making MIDCAB a more reliable and widely acceptable procedure.

Fig.9.2. Retractor and stabilizer for MIDCAB procedures.

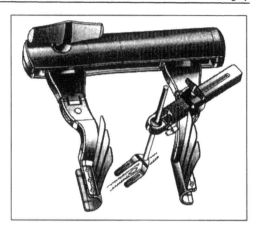

The patient who is a MIDCAB candidate generally has single vessel disease in the left anterior descending artery (LAD) distribution, although right coronary disease can also be bypassed with MIDCAB techniques. A small incision is made in the left anterior chest in the 4th intercostal space for LAD bypass. If necessary for exposure, the 4th costal cartilage may be excised. The left internal mammary artery is carefully identified and dissected cephalad using various commercially available exposure devices. The patient is moderately heparinized. The LAD is exposed, and vascular control achieved, generally with a proximally placed vessel loop positioned on a blunt needle. The stabilizer is positioned and the LAD opened. A vascular occluder may be inserted into the vessel. Alternatively, if the vessel is large and important, a vascular shunt may be placed to permit continued perfusion while maintaining vascular control and a bloodless field (Fig. 9.3). A carbon dioxide blower is helpful to remove residual blood. It is helpful for the blower to be attached to a gentle saline aerosolizing device to prevent dessication of the target tissues. The internal mammary is grafted to the LAD. The quality of the anastomosis is immeasurably improved with the stabilizing device, with which patencies (at least short term) generally approach standard LIMA to LAD anastomoses on the arrested heart.

A concept which has been proposed for multivessel disease is the so-called "hybrid" procedure. This combines a MIDCAB procedure with endovascular coronary procedures (angioplasty, stenting) either in a simultaneous or staged manner. The concept generally does not coincide with conventional teaching that surgical coronary revascularization should be as complete as possible to prevent future coronary events. Furthermore, subsequent reoperation of failed angioplasty attempts is made much more difficult in the presence of a patent LIMA graft. Proponents of the hybrid procedure contend that the procedure is useful as a salvage maneuver in high risk patients who might not tolerate a prolonged coronary revascularization procedure on pump. Furthermore, they contend that the most important anastomosis that determines long term patient survival is the LIMA to LAD anastomosis. The hybrid procedure may have some indications. However,

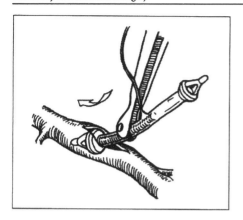

Fig. 9.3. Intracoronary shunt to achieve vascular control while maintaining perfusion.

these must be selected cases. The hybrid procedure should be used judiciously and is not to be considered the mainstream of coronary revascularization methods.

OPCAB

The MIDCAB technique does not address circumflex disease which is not easily accessible with a limited thoracotomy, and is, therefore, applicable to only a minority of CABG patients requiring complete surgical revascularization. "Off-pump coronary artery bypass" (OPCAB) refers to complete surgical coronary revascularization without the heart-lung machine. It is generally done via median sternotomy in order to approach the anterior surface of the heart as well as the less readily accessible areas through which the circumflex and right coronary vessels course. Exposure of all vessels is feasible with sternotomy, but even for single vessel LAD disease, a sternotomy approach may be preferred to MIDCAB. Compared to mini-thoracotomy, a sternal approach simplifies LIMA harvesting, identification of the LAD is done with greater confidence, and access for cannulation is readily available if emergency CPB is needed. Minithoracotomy does not necessarily result in less pain than a sternotomy, rather the converse is often true. However, one advantage of MIDCAB to OPCAB may be in reoperations where adhesion takedown through a sternotomy is obviated.

PATIENT SELECTION

Although most CABG patients tolerate CPB well, certain subgroups have a higher incidence of adverse outcomes from CPB, including the elderly and patients with prior stroke or severe pulmonary, renal, or ventricular dysfunction. It is these patients who may benefit the most with OPCAB compared to standard CPB. Ejection fraction should not necessarily be used as a basis to reject a patient from OPCAB. The "worst" operative candidate physiologically may paradoxically be the "ideal" patient for OPCAB.

OPCAB appears to have decreased morbidity compared to on-pump CABG, presumably because of elimination of CPB. Several studies trend toward a decrease in the incidence of stroke, renal failure, prolonged ventricular dependence and operative mortality. The effect appears to be even more pronounced in high risk patients.

Some groups of patients may anatomically be suboptimal OPCAB candidates. These include patients whose angiograms reveal small, intramyocardial, or heavily calcified vessels which may require endarterectomy. Cardiomegaly is a relative contraindication to OPCAB because these hearts are particularly difficult to position for OPAB, particularly in the circumflex distribution. Hemodynamically unstable patients are not candidates for OPCAB.

TECHNIQUE

After sternotomy and IMA harvesting, systemic heparinization is established using one-third to one-half of the CPB loading dose and the activated clotting time kept at 200-300 seconds. Reversal with protamine is done for most patients at the end of the procedure. Deep pericardial sutures are placed in strategic locations in the left posterior pericardium to facilitate pericardial retraction for cardiac elevation and exposure. Care is taken to avoid the phrenic nerve or pulmonary veins. Different amounts of traction on these sutures elevates the heart in different ways, by either lifting the apex up and out of the sternotomy or rolling the lateral heart toward the surgeon (Fig. 9.4). The Trendelenburg position further

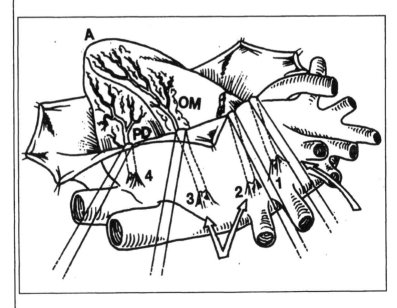

Fig. 9.4. Deep intrapericardial sutures which, when elevated, achieve exposure of the anterolateral aspect of the heart. Steep Trendelenburg achieves further exposure.

elevates the apex out of the chest. Since the deep pericardial suture adjacent to the left superior pulmonary vein could conceivably lacerate the left atrial appendage if pulled too taut, a rubber snare is used on these sutures to proved a softer buttress.

One of the several commercially available stabilizing systems is used to immobilize the area of the target vessel. Vascular control, shunts, and occluders are used in a manner similar to that described for MIDCAB. As a general rule, the first vessel grafted is the LAD because of its ease and importance (Fig. 9.5). The subsequent order of vessel grafting depends on the size, importance, and ease in maintaining tissue perfusion while achieving exposure. Generally, vessels with the least difficult exposure and technical grafting complexity are done first. The greatest danger to injury to the LIMA to LAD anastomosis comes from torquing the heart inferiorly and to the right, such as when exposing the ramus intermedius or high obtuse marginals. Hence, firmly anchoring the LIMA pedicle to the epicardium is essential. All distal anastomoses may be completed first, followed by proximal grafting to the aorta using a partial occlusion clamp. Alternatively, proximal grafting may be accomplished in turn after each distal.

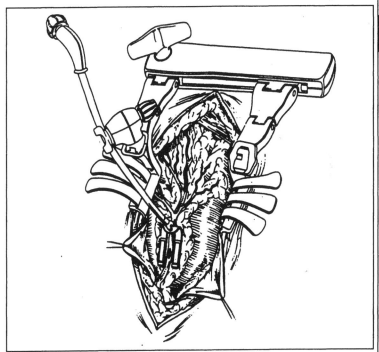

9

Fig. 9.5. Retractor and stabilizer positioned via median sternotomy for OPCAB procedures. The LAD vessel is being exposed.

Cardiac manipulation to achieve exposure may be poorly tolerated hemodynamically. This is especially true for exposure of the obtuse marginal vessels. Active, vigilant participation by the anesthesiologist using anticipatory, as well as reactive, strategies is a requirement for safe and successful OPCAB. Short-acting beta blockers such as esmolol help control the heart rate and reduce cardiac oxygen consumption. Phenylephrine, dopamine, and epinephrine support vascular resistance and perfusion pressure, and volume loading with infusions and Trendelenburg position optimize cardiac output during cardiac manipulations.

An important risk of hemodynamic instability arises when achieving vascular control of the right coronary artery. Intraoperative cardiac arrest may arise in this instance from ventricular tachycardia, fibrillation or bradydysrhythmias. If possible, OPCAB grafting of the branches of the right coronary artery, rather than the right coronary itself, may be preferable to help minimize this risk. If grafting on the right coronary directly is necessary, a shunt occluder which maintains perfusion may be of the most benefit. Although extubation in the operating room is possible, it is not necessary and OPCAB patients can usually undergo early extubation within several hours postoperatively.

Probably the most popular and easy to use device is the Medtronic Octopus suction-stabilizing device to immobilize the target area. This is often used in conjunction with a suction cup over the ventricular apex to lift and position the heart, primarily to achieve exposure of the inferior and lateral aspects of the heart. This generally results in less hemodynamic compromise than if sponges or other retraction techniques are used. The apical suction cup obviates the need for deep pericardial retraction sutures.

THORACOTOMY FOR OBTUSE MARGINAL OPCAB

Reoperative coronary surgery in patients with patent IMA grafts may be hazardous. If such a patient only requires grafting of the circumflex system, a thoracotomy approach with single-lung ventilation may be useful to avoid cardiac injury from sternal reentry. This thoracotomy approach (usually 4th interspace) may be combined with OPCAB instrumentation to provide local stabilization for obtuse marginal grafting (Fig. 9.6). The proximal anastomosis should be done to the descending thoracic aorta or, alternatively, the splenic or subclavian arteries in the case of a heavily calcified descending thoracic aorta. It is important to keep the pelvis corkscrewed and the left groin in clear view in case femoral-femoral CPB becomes necessary. Preoperative lung function tests to ensure that the patient can tolerate a thoracotomy is important. Off-pump thoracotomy grafting of the obtuse marginal targets is technically more difficult than other MIDCAB or OPCAB grafting. This relates to the distance of the lateral heart vessels from the lateral chest wall, resulting in a deeper hole within which the surgeon must maneuver, and less steady stabilization by the foot plate.

In summary, OPCAB, by sternotomy or thoracotomy, should not be done if it cannot be done safely or if the quality of the anastomoses is felt to be inferior to

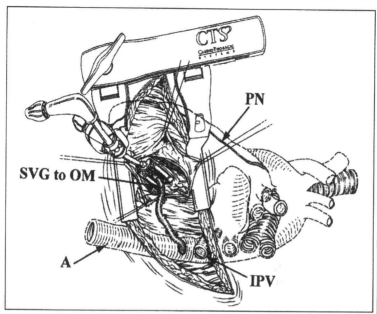

Fig. 9.6. Retractor and stabilizer positioned via a left posterolateral thoracotomy for off-pump grafting of the marginal artery in a redo setting.

those performed on-pump. It requires a level of intensity and cooperation from both surgeon and anesthesiologist that differs from most routine pump cases. Long-term graft patency with OPCAB using stabilizing devices is not yet available, but early results appear encouraging.

It is important, as with any new procedure, to ensure that reason and sound ethical principles apply. This has not necessarily been followed in the past 30 years of modern medicine in matters of life, death, faith and morals. In the case of cardiothoracic surgery, historically steep "learning curves" when there were no alternative treatments, were acceptable, as it was ethically not only justified, but necessary to relieve human suffering. The same type of learning curve is not acceptable today. "Learning curves" can justify some surgical experiences, but it is our duty to ensure that the end is worth the means required to achieve it.

SUGGESTED READING

1. Roach GW, Kandruger M, Mangano CM et al. Adverse cerebral outcome after coronary bypass surgery. NEJM 1996; 335:1857-1863.
2. Svensson LG, D'Agostino RS. Minimal-access aortic and valvular operations, Including the "J/j" incision. Ann Thorac Surg 1998; 66:431-5.
3. Buffolo E, de Andrade JCS, Branco JNR et al. Coronary artery bypass grafting without cardiopulmonary bypass. Ann Thorac Surg 1996; 61:63-6.

4.   Pfister AJ, Zaki MS, Garcia JM et al. Coronary artery bypass without cardiopul-
      monary bypass. Ann Thorac Surg 1992; 54:1083-92.

5.   Calafiore AM, Teodori G, DiGiammarco G et al. Minimally invasive coronary
      artery bypass grafting on a beating heart. Ann Thorac Surg 1997; 63:S 72-5.

6.   Subramanian VA. Less invasive arterial CABG on a beating heart. Ann Thorac
      Surg 1997; 63:S 68-71.

7.   Baumgartner FJ, Gheissari A, Capouga ER et al. Technical aspects of total
      revascularization in off-pump coronary bypass via sternotomy approach. Ann
      Thorac Surg 1999; 67:1653-8

8.   Yokoyama T, Baumgartner FJ, Gheissari A et al. Off-pump vs. on-pump coro-
      nary bypass in high-risk subgroups. Ann Thorac Surg 2000; 70:1546-50

9.   Baumgartner FJ, Yokoyama T, Gheissari A et al. Effect of off-pump coronary
      bypass on morbidity. Am J Cardiol 2000; 86:1021-2

10.   Stamou SC, Jablonski KA, Pfister AJ et al. Stroke after conventional vs. minimally
      invasive coronary bypass.

ICU Management

Fritz J. Baumgartner, John R. Robertson, Bassam O. Omari

PHYSIOLOGY

The most important hemodynamic parameter that must be estimated is the adequacy of oxygen delivery to the tissues. This can be assessed in several ways: blood pressure, heart rate, urine output, warm, dry extremities (vs. cold, clammy extremities), and capillary refill (greater than 3 seconds is considered abnormal). These are the old-fashioned, time-honored methods of assessing the adequacy of tissue perfusion which every physician must be skilled at. Frequently, a good assessment of a patient's condition can be made by these parameters and the diagnosis can be fine-tuned using more specific parameters which can be obtained from the Swan-Ganz catheter and central venous pressure monitoring lines.

Shock can be defined as a condition of inadequate tissue perfusion and is due to one of four possible etiologies: (1) cardiogenic shock; (2) hypovolemic shock; (3) septic shock; and (4) neurogenic shock. Another type of shock, tamponade, is actually a subset of cardiogenic shock although it is not a dysfunction intrinsic to the myocardium but rather an external force acting on the myocardium. The Swan-Ganz catheter may be a valuable aid in differentiating these types of shock. This can be done by an evaluation of the cardiac output, cardiac index, systemic vascular resistance, pulmonary vascular resistance, CVP, pulmonary artery pressure, and pulmonary capillary wedge pressure (from which can be estimated the left atrial pressure and the left ventricular end diastolic pressure).

The mixed venous oxygen saturation is also a useful parameter in distinguishing the types of shock. The cardiac output is the stroke volume times the heart rate and is described in liters per minute. The cardiac index is the cardiac output divided by the surface area in meters squared. Normal cardiac output is in the range of 4-5 liters per minute. The cardiac index is in the range of 2-3 liters per minute per meters squared. The "filling pressures" is a term frequently used on the Cardiothoracic Surgery service and refers to the pressure within the chambers of the heart. This can be estimated by the CVP which mirrors the right atrial filling pressure and is normally in the range of 2-5 mmHg. The right ventricular pressure is more difficult to measure and is in the range of 30/0 mmHg. The pulmonary artery pressure is in the range of 30/10 mmHg and is measured by the Swan-Ganz distal port. The left atrial pressure is in the range of 8-10 mmHg and can be estimated by the pulmonary capillary wedge pressure. When the distal end of the pulmonary artery catheter is wedged in the pulmonary artery, it reflects the pressure in the pulmonary veins, which in turn reflects the pressure in the left atrium.

The left ventricular pressure is in the range of 120/5 mmHg. The left ventricular end diastolic pressure can be estimated by the left atrial pressure, which in turn can be estimated by the pulmonary capillary wedge pressure.

The systemic vascular resistance is an important parameter. It is one of the primary factors related to afterload. Elevated systemic vascular resistance may occur in the presence of hypovolemic shock where catecholamine surge causes profound vasoconstriction and elevation of the systemic vascular resistance to try to support the blood pressure. In certain instances in the postoperative cardiac surgery patient, pharmacologically decreasing the systemic vascular resistance may be very helpful in improving cardiac output since the heart will have to contract against less resistance. Of course if the problem is volume-related peripheral vasoconstriction, then simply volume loading the patient will decrease the systemic vascular resistance physiologically rather than pharmacologically.

The systemic vascular resistance can be calculated by the following formula: $[(MAP-CVP)/C.O.]$ x 80 = dynes cm^{-5}. The mean arterial pressure can be calculated as two times the diastolic plus the systolic all divided by three. If one does not use the conversion factor 80, then the value obtained is in Woods units, which was used in the past more frequently. Pulmonary vascular resistance is an assessment of the resistance in the pulmonary vasculature. This is calculated as: $[(MPP-wedge)/C.O.]$ x 80. MPP is the the mean pulmonary artery pressure and is calculated as is MAP. This is more commonly expressed in Woods units than the systemic vascular resistance is.

The filling pressures may have a direct role in cardiac output according to the Starling curve. Initially cardiac output goes up dramatically with increasing left atrial pressure; however, at a critical point the cardiac output starts to plateau and when the pressure is so high that distention occurs, then cardiac output falls. The improvement of cardiac output with increasing left atrial pressure is less pronounced with ischemic or compromised hearts. In tamponade, volume loading is one of the key therapeutic modalities available prior to surgery.

Inotropes shift the Starling curve in favor of improved cardiac output with relation to filling pressures; this is done by improving contractility. Unloading the ventricle by decreasing afterload also improves cardiac output. The Swan-Ganz catheter is an extremely useful device for measuring filling pressures of the right atrium, pulmonary artery pressure, pulmonary capillary wedge pressure, cardiac output, systemic vascular resistance and pulmonary vascular resistance. It has three ports. One is a proximal infusion port, the other is a proximal pressure port, and the third is a PA distal pressure port. There also exists a balloon inflation device and a thermistor for measuring cardiac output via the thermodilution method.

When placing the Swan-Ganz, there is a typical pressure trace as the balloon transverses the right atrium, the right ventricle, the pulmonary artery, and then is wedged into position. Typically, the balloon is inflated when the catheter is inserted 15 cm and is then passed through the right heart into the pulmonary artery and wedged into position. Normally, it wedges in the right pulmonary artery although it may go into the left pulmonary artery on occasion. It is important to keep the balloon inflated when directing the catheter through the right heart lest it bang too forcefully against the right heart endocardium, possibly causing perforation or becoming entrapped in the trabeculations.

Right bundle branch block is a frequent finding during passage of the catheter and one must be extremely cautious, especially in the presence of left bundle branch block lest a complete heart block occur. PVCs and ventricular tachycardia are also common.

Another important parameter that the Swan-Ganz catheter can evaluate is the mixed venous saturation (SVO_2). This is the saturation of the venous blood retrieved from the pulmonary artery. Recall that the PvO_2 in the pulmonary artery is about 40 mmHg which corresponds to 75% saturation. This is normal, i.e., 25% less than the normal PaO_2 which is 100%. An adequate mixed venous saturation is usually in the range of 25% less than the arterial saturation. One may recall that the most desaturated blood in the body is drained from the coronary sinus because of the high oxygen use by the normal heart. Thus venous saturation of the coronary sinus is less than in any other part of the body. Venous saturation in the inferior vena cava is fairly high because the kidneys take up or utilize less oxygen than other organs. The superior vena cava generally has a lower mixed venous saturation than the inferior cava because the brain utilizes more oxygen. These sources of venous blood "mix" in the right atrium and then go through the ventricle up into the pulmonary artery. This is why samples taken through the distal tip of the pulmonary artery catheter are called mixed venous samples.

The mixed venous oxygen saturation is an important hemodynamic parameter which can be used to define trends in a patient's hemodynamic status and clinical outcome. For example, patients with impending tamponade will frequently show a decrease in their mixed venous oxygen saturation long before changes in heart rate or blood pressure occur. The reason is that the mixed venous condition indicates poor tissue oxygenation resulting from poor cardiac output related to the tamponade. This may not yet be reflected by the blood pressure or heart rate. A similar situation occurs in a failing myocardium from poor contractility secondary to ischemia, edema or other causes.

There are only three factors which will cause decreased mixed venous saturation and should be known by every surgery resident managing patients in the CSICU. These include:

1. *Decreased cardiac output:* This bespeaks of low tissue perfusion resulting in increased extraction of oxygen from the available blood flow, resulting in decreased mixed venous oxygen saturation. This results from poor inotropy, low volume or slow heart rate. It can be confirmed by other hemodynamic parameters including cardiac output. Management involves inotropes, volume or increasing the heart rate.

2. *Decrease in oxygen content of the blood:* The formula for oxygen content is (Hg) x 1.38 x % saturation. As one can see from the formula, oxygen content can be decreased by both decreased hemoglobin concentration, i.e., anemia or low hematocrit, or can be effected by the saturation of the blood, as for example, the patient who is not being ventilated with

adequate FIO_2 during the initial perioperative period. Either of these will decrease the oxygen content and, thus, the saturation of the venous blood. This occurs because of increased extraction from the available oxygenated red blood cells by the oxygen hungry tissues.

3. *Increased metabolic demand by the peripheral tissues*: This occurs classically in cases of shivering in the immediate postoperative period. One often sees a patient shivering simultaneous with the mixed venous saturation dropping precipitously. The problem can be at least partially corrected by paralyzing the patient to prevent the shivering reaction.

Thus in all the major types of shock seen in the immediate postoperative period after cardiac surgery, the SVO_2 will drop. One particular instance, however, of shock in which the SVO_2 does not decrease is sepsis. This is a very important point in differentiating sepsis from other conditions of shock. In conditions of sepsis, AV shunts are opened in which oxygenated blood bypasses the tissues and goes directly into the venous circuit resulting in an increase in SVO_2.

Table 10.1 shows the differentiation of hypovolemic, cardiogenic, and septic shock based on the blood pressure, CVP, wedge pressure, cardiac output, SVO_2 and the SVR. Additionally, tamponade is shown which basically mimics cardiogenic shock with the exception that the filling pressures are even more markedly elevated with tamponade than in the latter and tend to equilibrate at a higher level.

Pulmonary embolus is also shown in Table 10.1. This is a rare event after cardiopulmonary bypass; however, it does occur as early as a week postoperatively. Since the patient has been fully systemically heparinized, it is extremely rare to develop a pulmonary embolus in the immediate perioperative period. However, a week postoperatively, we personally have seen several patients develop massive pulmonary emboli with a high mortality. It is not a condition to be taken lightly. Pulmonary emboli too will decrease the SVO_2 because of inadequate oxygenation of the available hemoglobin. The patient dies after massive pulmonary embolus for two reasons: asphyxiation from inadequate blood oxygenation and decreased cardiac output with poor volume loading of left heart because of obstruction of the pulmonary artery.

Cardiac output measurements can be made by several methods. These include:

1. Indicator dilution method with indocyanine green.
2. Thermodilution technique with the Swan-Ganz catheter.

Table 10.1. Hemodynamic parameters related to type of shock

Type of shock	BP	P	CVP	Wedge	C.O.	SVO₂	SVR
Hypovolemic	↓	↑	↓	↓	↓	↓	↑
Cardiogenic	↓	↑	↑	↑	↓	↓	↑
Septic	↓	↑	↓	↓	↑↑	↑	↓↓
Tamponade	↓	↑	↑↑	↑↑	↓	↓	↑
PE	↓	↑	↑	−	↓−	↓	↑

3. Fick method which is basically evaluation of the rate of oxygen consumption. This is a function of the rate of blood flow times the rate of oxygen pickup by the red blood cells. Hence, cardiac output equals oxygen consumption divided by AV oxygen difference times ten.

POSTOPERATIVE PROTOCOLS AT HARBOR-UCLA

POSTOPERATIVE NOTE
This should be written by the junior resident accompanying the patient from the operating room or the junior resident on call.
It should include:
- preoperative diagnosis
- postoperative diagnosis
- operation performed, including the number, type and positions of all coronary bypass grafts; the size, type and position of any graft or prosthetic material
- the operative team
- the type of anesthesia
- a list of all lines, catheters and tubes
- the operative findings
- a list of all IV drips and their present infusion rates
- any intraoperative or immediate postoperative complications

GENERAL PRINCIPLES FOR INITIAL CARE ON ARRIVAL TO 3W CTU
Vital Signs
Record all vital signs immediately on arrival.
Monitoring Lines
Detach and reattach to the 3W CTU monitors, one line at a time. Never disconnect and change all lines simultaneously.
Stat Labs
Chem-7, CBC with platelet count, PT, PTT, Chest XR, EKG.
Chest Tube Drainage
Record the chest tube drainage in the Pleur-Evac on arrival to the 3W CTU.
Ventilation
Patients returning from the operating room on a ventilator with an endotracheal tube in place should have ventilator settings determined jointly with the anesthetist caring for the patient. The appropriate FIO_2, tidal volume calculated on the basis of 10-15 cc/kg and ventilatory rate should be determined at this time. Changes in these settings are made by the surgical resident based upon clinical and laboratory findings. It is important to consult the appropriate respiratory therapist caring for any patient when any ventilatory changes are made. This is mandatory since these individuals are highly skilled in artificial ventilation techniques and will help prevent errors in management. Patients on ventilators are at

high risk of inappropriate ventilation and therefore clinical examinations should be performed frequently. The resident must assess the appropriate rise and fall of the individual patient's chest, as well as auscultate both lung fields to ensure ventilation bilaterally. Examination of the position of the endotracheal tube both clinically examination and on the initial postoperative chest x-ray is mandatory. In addition, the surgical house officer is expected to secure the endotracheal tube. Arterial blood gas determination should be performed whenever any question exists about the ventilatory status. However, blood gas determination should not delay making a critical decision in management. If one suspects that either ventilation or oxygenation is inappropriate, steps should be taken to adjust the situation prior to obtaining the actual blood gas result. Blood gas determinations are expensive and should only be ordered when absolutely indicated.

NG Tube

All postoperative cardiac surgical patients are allowed nothing by mouth in the immediate postoperative period and should return to the unit with a nasogastric tube in place. These tubes are necessary to prevent the complications of gastric dilatation and aspiration. It is the house officer's responsibility to ensure that these tubes are functioning appropriately. All tubes used are the Salem sump variety, and therefore the sump portion of the tube must be working correctly. Frequent irrigations with small volumes of saline may be required to ensure appropriate function of these devices. Correct positioning of these tubes should be ensured on the postoperative chest x-ray.

Chest Tubes

All patients return from the operating room with anywhere from one to three chest tubes in place. It is routine for our patients to receive an anterior and posterior mediastinal chest tube; however, with dissection into either the right or the left hemithorax or in cases where the internal mammary was taken down, an additional chest tube may be placed in the respective hemithorax. It is the resident's responsibility to inspect these tubes and to ensure that all junctions are air tight. The chest tubes should be connected to a Pleur-Evac suction apparatus with -20 cm of water suction. These tubes should be inspected frequently to determine the rate of postoperative bleeding and air leak. No patient should ever have his or her chest tube opened under any condition without contacting the Chief Resident or attending physician first. The Pleur-Evac system is a closed system and therefore opening it violates surgical sterile technique. Any clots appearing in the chest tubes themselves may be drained by gently tapping the tube with a metal object or one's hand. Routine chest tube stripping is not allowed.

Foley Catheter

All patients return with a Foley catheter in place. It is the resident's responsibility to ensure that this system is draining adequately. The Foley catheter itself should be connected to a urinometer measuring system so that accurate hourly urine output can be obtained. Remember, urine output is a very sensitive indicator of cardiac output. In addition, hourly specific gravity should be obtained to further aid in management. The Foley catheter system is a closed sterile system and therefore should not be opened under any circumstances. Catheters which are not functioning appropriately should be removed and replaced with an entirely new system.

Arterial Line

All patients will return with an arterial line in place. This should be inspected to ensure that appropriate dressing and antibiotic ointment has been applied to the puncture site. The wave tracing and transducer performance should also be inspected. Care should be taken to ensure that no air resides in any portion of the closed transducer system. The arterial line is usually discontinued within the first 24-48 hours postoperatively.

Swan-Ganz Catheters

Most adult patients return with Swan-Ganz catheters in place. These devices should be inspected to ensure that the point of entry into the skin is appropriately covered with antibiotic ointment and dressing. Patients should be tested. Appropriate wave tracings should be analyzed.

The device is used for three primary purposes:

1. Determine pressure measurements (RA, PA, pulmonary capillary wedge).
2. Determine cardiac output (thermodilution technique).
3. Determine the mixed venous O_2 saturation. When determining pressures, care should be taken to ensure that the transducer functions.

When analyzing cardiac output, appropriate volumes and temperatures of injectables should be checked along with the appropriate computer constant entered in the thermodilution computer. When obtaining mixed venous O_2 saturations, the distal port of the Swan should be used to obtain pulmonary capillary wedge pressures; this is not done without direct permission from either the Chief Resident or attending physician. In most cases the pulmonary artery diastolic pressure is an accurate representation of the pulmonary capillary wedge pressure. We must always be aware of the possibility of permanently wedged catheters and when this occurs, the resident must free the catheter.

Left Atrial or Right Atrial Lines

Occasionally adult cardiac patients return with one or both of these lines in place. Care to the puncture site in the skin should be maintained as well as appropriate evaluation of the pressure monitoring system. Specific attention should be paid to the left atrial line. At no time must any air bubbles be injected into this system for they will enter the left side of the heart and potentially cause significant, if not fatal, emboli.

Pacemaker Wires

A majority of our patients return with several pacemaker wires in place. These wires are carefully positioned so that the ventricular leads are to the patient's left and the atrial leads to his right. Ground leads can be identified by the wiring to the skin. Residents must be familiar with the positions of these leads on each patient and know how to connect them to the appropriate generator box if the situation is called for. Patients who return from the operating room where evidence of heart block and/or bradycardic rhythms occurred should be left attached to the pacer generator with a demand mode backup.

EKG Monitoring Pads

Interruption in EKG electrical activity should promote a search for a loose electrical connection or EKG pad. This should be recognized and corrected immediately.

Electrocardiogram

An electrocardiogram should be done shortly after the patient arrives in the intensive care unit. This EKG should be evaluated immediately and should remain the baseline for evaluation of future EKGs and for evaluation of any arrhythmias or conduction disturbances which may be present. An EKG should be ordered on a QAM basis for the first 3 postoperative days.

Chest X-Ray

A chest radiograph should be obtained shortly (<1 hr) after the patient arrives in the CTU. The purpose of this film is to ascertain the position of all intrathoracic tubes (CVP, LA, Swan-Ganz catheter, endotracheal tube, chest tubes, nasogastric tube). It also gives information about the presence of air and blood in the mediastinum and pleural spaces. It is a baseline of heart size and hence the appropriate x-ray settings and distance used to shoot these films should be written at the foot of each bed and all subsequent x-rays done using the same technique.

LOW OUTPUT STATE

General Principles

The most common, yet most serious, problem in the postoperative period is low output state. It is well known that a drop in the cardiac index below 2 l/min/m^2 is directly associated with increased mortality. It is therefore mandatory that surgical residents caring for postoperative cardiac patients become adept in early recognition and treatment for low cardiac output.

Recognition

The bedside diagnosis of low cardiac output can be divided into clinical and measured parameters.

The clinical diagnosis of low cardiac output resolves around signs and symptoms of organ hypoperfusion. The primary systems which we use to evaluate this state are the skin, the kidney and the brain. When cardiac output is low, skin perfusion is dramatically reduced. Therefore, the skin becomes cold and clammy. In addition, a drop in cardiac output significantly reduces kidney perfusion with the resultant rapid decrease in urine output. And finally, low cardiac output dramatically affects cerebral perfusion with resultant decreased consciousness.

Direct measurements of cardiac output can be obtained by the following: utilizing the Swan-Ganz catheter and the thermodilution computer. In addition arterial blood gas analysis showing the development of metabolic acidosis should alert one to the potential of low output state with resultant lactic acidosis. Finally, analysis of blood obtained from the distal port of the Swan-Ganz catheter showing a low mixed venous oxygen saturation should again suggest the diagnosis of low output state.

Treatment

The treatment of a low cardiac output revolves around the understanding of the variables controlling cardiac output. Mathematically, cardiac output is equal to cardiac stroke volume times heart rate. Stroke volume itself is determined by preload, afterload and myocardial contractility. Hence manipulation of the patient's heart rate, his preload, his afterload or his myocardial contractility will allow one to adjust the overall cardiac output.

Problems in Heart Rate and Rhythm

It is not uncommon for postoperative patients to experience problems with significant bradycardia. In addition, many patients, particularly those with surgery around the tricuspid valve, have problems with heart block. In both of these conditions the treatment of choice is atrial, ventricular, or atrial-ventricular pacing. All postop cardiac patients have either ventricular, atrial or atrial and ventricular pacing wires in place. Therefore the initial treatment for these conditions is external pacing. In addition, patients who are ventricularly paced can have a drop in cardiac output due to the fact that they have lost their atrial component of ventricular filling. Therefore, switching these individuals to atrial-ventricular pacing may significantly improve their cardiac output. In those individuals where pacing wires are either not present or have been removed, or are not capturing, pharmacologic manipulation of heart rate is indicated. The primary drugs of choice used are atropine and isoproterenol. Finally, in those individuals where a particular dysrhythmia adversely affects the cardiac output, treatment (see section on dysrhythmias) of the underlying dysrhythmia may significantly improve the deteriorating output state.

Treatment of Inadequate Preload

Preload may be defined as the ventricular fiber length at the end of diastole. Hence preload is closely related to the ventricular end diastolic pressure which can be approximated by measurement of the pulmonary capillary wedge pressures or the pulmonary artery diastolic pressure. Problems in preload arise when one or two conditions develops. The first is hypovolemia. Hypovolemia occurs due to blood loss, plasma loss, significant venodilation, or to inadequate volume replacement. Under these conditions, the appropriate treatment would be to replace the missing component. In conditions where significant blood loss has occurred and the hematocrit demonstrates a loss of red cell mass, transfusions of packed red blood cells would be appropriate. In conditions where significant plasma loss has occurred the hematocrit itself may be high but the pulmonary capillary wedge pressure would be low demonstrating a depleted intravascular space. Under these conditions, fluid replacement is in order. On our service it is preferred that a colloid-type agent be used so that the replacement remains in the intravascular space. It is our preference to use 5% albumin or Hespan when volume replacement is required.

The other problem with preload is impaired filling. Impaired ventricular filling occurs when conditions such as cardiac tamponade, tension pneumothorax, or excessive positive pressure breathing interferes with the normal return of blood to the heart. The treatment of impaired ventricular filling revolves around the elimination of the cause of impairment. In the case of tamponade a return to the operating room to drain the mediastinum would be appropriate. In the presence of tension pneumothorax a chest tube will quickly relieve the problem. Finally, in situations where positive pressure breathing may be interfering with blood return, a diminution of the end expiratory pressure may be necessary.

Treatment of Problems in Afterload

Afterload may be defined as the pressure that the heart must overcome in order to open the aortic valve successfully and eject its ventricular volume. More simply, it is the aortic impedance to ventricular ejection. Problems develop in afterload when either the systemic vascular or pulmonary vascular resistances are either increased or decreased. An increase in afterload usually occurs post-bypass and when the patient experiences significant pain. This is due to excessive circulating catecholamine levels produced by the bypass technique or by postoperative pain. Increased pulmonary vascular resistance may occur secondary to chronic pulmonary vascular changes or secondary to a pulmonary embolus. The treatment of increased afterload therefore revolves around the use of vasodilatory drugs. The drug of choice used on our service is nitroprusside. The drug is given by continuous IV infusion, with close observation of blood pressure and cardiac output. It is our first choice because it has an immediate onset of action and its effects are quickly terminated upon discontinuation of infusion. In situations where afterload is significantly decreased secondary to marked vasodilation due to either drugs or sepsis, the treatment is vasoconstriction of the distal arteriolar beds. Drugs such as neo-synephrine may be utilized for this purpose. In conditions where sepsis is the underlying cause, treatment of the septic process remains the primary means of correcting the afterload situation.

Treatment of Inadequate Myocardial Contractility

Myocardial contractility is defined as the inotropic state or contractile state of the heart. Mathematically, it is the velocity of ventricular contraction when afterload and preload are kept constant. Problems in postoperative contractility occur when patients have had:

1. pre-existing ventricular dysfunction
2. inadequate intraoperative myocardial protection
3. coronary air or particulate embolism
4. residual, uncorrected cardiac defects
5. acute myocardial infarction
6. pharmacologic myocardial depression
7. hypoxemia
8. acid base imbalance
9. septic or prolonged hypovolemic shock
10. fever or tachycardia
11. allergic reaction
12. hypocalcemia

Treatment of contractility is inotropic stimulation to the heart. It must be remembered that when problems in cardiac output occur that we manipulate first heart rate, second the preload, third the afterload and finally, and only then, the contractility. Evidence of myocardial contractile failure occurs when all other parameters have been correctly stabilized yet cardiac output remains low and ventricular end diastolic filling pressures or pulmonary capillary wedge pressures remain high. Then inotropic drugs should be given. The primary drug used for this purpose on our service is dopamine. Dopamine is given as an IV infusion, starting at around 2 to 5 mcg/kg/min. It may be increased as needed. In addition, Dobutamine may also be utilized as a primary agent or to supplement a patient already on dopamine. Isoproterenol, another very potent inotropic drug, may also be utilized in patients with failing myocardial contractile function.

CARDIAC CARE
Hypotension
This discussion will define hypotension as the systolic blood pressure less than 90 mmHg. It is recognized that some patients have a blood pressure which normally runs around 90 systolic, and in these cases nothing need be done. Also, in elderly patients blood pressures much higher than 90 may be required. However, for the average patient a blood pressure of less than 90 systolic in combination with signs and symptoms of tissue hypoperfusion or low cardiac output requires intervention. The causes of hypotension are many and must each be recognized and treated independently.

Hypovolemia
The hypovolemic patient presents with low central venous pressure and low pulmonary capillary wedge pressure. This individual needs expansion of his intravascular volume. This may be done with colloid-type agents. If the hematocrit is low, replacement with packed blood cells is the treatment of choice. When the hematocrit is high and the circulating intravascular volume low, replacement with agents such as 5% albumin or Hespan become the treatment of choice. Immediately postop and during the first few postoperative days patients have significant third space fluid accumulations. Therefore, mannitol or 25% albumin, which shifts fluid from the third space to the intravascular space, is frequently used on our service.

Perioperative Myocardial Infarction
Diagnosis of postoperative myocardial infarction can usually be made with EKG changes, increases in the left ventricular end diastolic filling pressures, decrease in cardiac output, and an increase in the CPK MB fraction. The standard treatment for a myocardial infarction should be initiated.

Dysrhythmias
Various dysrhythmias seen in the postoperative course can lead to problems of hypotension. The treatment of the particular dysrhythmia causing the problem should therefore be initiated (see section on dysrhythmias for appropriate management).

Massive Transfusion of Blood Products

It is well recognized that massive transfusion of blood products is complicated by excessive unbound citrate contained within the blood product to prevent co-agulation. The citrate abruptly decreases the ionized circulating calcium level and can lead to severe myocardial depression. Therefore, whenever blood products are given in massive rapid infusions, it is important to give calcium chloride via central vein.

Medications

Various medications can cause postoperative hypotension. It is not uncommon for the infusion of nitroglycerin or nitroprusside or the oral administration of inderal or nifedipine to cause significant hypotension. It is important to recognize these agents as the cause and to discontinue their use as indicated.

Tension Pneumothorax

A tension pneumothorax may develop at anytime in the postoperative period. It may occur secondary to injury to the lung at the time of surgery or to rupture of a pulmonary bleb due to excessive positive end expiratory pressures. It is important to recognize this problem and to treat it with placement of a chest tube.

Hypertension

Generally we like to keep our patient's systolic blood pressures below 120-130 mmHg. Blood pressures above this level for the average cardiac patient are excessively high. It is not uncommon for a patient who is not bleeding postoperatively to suddenly start bleeding from suture lines when the blood pressure is elevated. In addition, the work required by the recovering postoperative heart is markedly increased with systolic hypertension. The causes of postoperative hypertension are several.

Post-Pump State

All patients who have been on cardiopulmonary bypass experience an obligatory increase in circulating catecholamines. This results in increased myocardial contractility and peripheral vasoconstriction. This combination is a set up for postoperative blood pressure elevation. The treatment of choice therefore is the immediate institution of nitroprusside drip titrating the dose of medication by the systolic blood pressure.

Pain

Pain, like the post-pump state, produces an excessive level of circulating catecholamines with a commissurate increase in contractility and increase in peripheral vascular resistance. The treatment of this problem revolves around the IV administration of morphine sulphate between 2-10 mg IV push Q 3-4 hours to control postoperative pain.

Rapid Reversal of Narcotics

Most of our patients are on narcotics and occasionally the utilization of narcan to reverse the effects of these drugs for either respiratory or other reasons may occur. One must be cautious whenever giving antianalgesic medications for they can produce an abrupt rise in blood pressure to greater than 220 mmHg and disrupt both the aortic suture line and/or pre-existing arterial defects, such as cerebral aneurysms. Use antianalgesic medications only when absolutely indicated.

Post-Coarctation Repair

It is well known that post-coarctation repair individuals frequently develop a severe hypertensive response. This may be accompanied by abdominal ischemia as well as disruption of various suture lines. Therefore, frequent examinations of the abdomen as well as the institution of nitroprusside and/or esmolol to control the paradoxical hypertension should be initiated. The hypertension associated with this disease state may be prolonged and these patients may need to be switched to oral medications such as enalapril.

Post-Aortic Aneurysm Repair

This is a condition provoked by long-standing essential hypertension. Therefore, in the postoperative period, these individuals usually have sustained systolic hypertension. Again, it is mandatory to control these individual's pressure with drugs such as nitroprusside or esmolol.

Risks of allowing the blood pressure to get out of control with a new aortic prosthesis in place are massive postoperative hemorrhage and occasional disruption of the entire suture line and death.

Dysrhythmias

Sinus Tachycardia

The therapy of this condition revolves around treating the underlying cause. Under special circumstances where the rapid heart rate may interfere with cardiac output, esmolol given IV or metoprolol PO may be utilized to control the tachycardia response.

Atrial Premature Contractions

This dysrhythmia requires no treatment; however, if the premature contractions are very frequent, they may be early indications of an atrium in which atrial flutter or fibrillation will occur. Under these circumstances, antiarrhythmic drugs such as digoxin may be started. In addition, the prophylactic use of metoprolol may prevent these arrhythmias (see section on special medications for explanation).

Atrial Flutter/Fibrillation

No treatment is required if the ventricular rate is reasonable. However, if the rate is fast, one should first attempt to control the rapid ventricular rate with IV digoxin. Usually, push IV digoxin to a point where the rhythm comes under control. For more rapid control of the arrhythmia, IV verapamil can in approximately 20-25% of the cases control the arrhythmia. In addition, IV diltiazem has been used to slow the rapid rate of contraction. In cases where an immediate response is absolutely necessary, i.e., significant problems in low output state, D.C. synchronized countershock may be utilized. Finally, rapid atrial pacing requiring a special external pacemaker box (a box with the capability of firing greater than 600 times per minute) can occasionally control this rhythm disturbance.

Paroxysmal Atrial Tachycardia

PAT is a frequent disturbing dysrhythmia. We use prophylactic PO metoprolol in an attempt to control this problem. However, patients that develop this rhythm disturbance can have a rapid deterioration in cardiac output. The drug of choice for control of this rhythm disturbance is verapamil. An IV push of between 0.075 and 0.15 mg/kg usually will bring this rhythm disturbance under control. Other medications which can be used to control PAT are IV digoxin or metoprolol. Cardioversion can also be used to control the rhythm disturbance, however, since the advent of verapamil this is seldom necessary.

Sinus Bradycardia

This condition can easily be treated with either atrial or ventricular pacing. All postop cardiac patients have pacing wires and hence, connection of the wires to an external pacing box can result in immediate control of the rhythm problem. Where the wires are not present or where they are not functional, IV push of atropine or continuous infusion of isoproterenol may relieve the rhythm disturbance.

Nodal Rhythm

No treatment is required if the rate is reasonable and blood pressure is satisfactory. However, if the rate drops and cardiac output is diminished, pacing, either atrial or ventricular, becomes the treatment of choice. Occasionally, where both atrial and ventricular wires are in place, AV sequential pacing may be utilized so as to gain the added improvement in cardiac output by allowing the atrial kick. In situations where pacing is not available, IV infusion of isoproterenol may be utilized.

Premature Ventricular Contractions

A few premature ventricular contractions need only be watched; however, frequent premature contractions (greater than 5/min), those occurring in runs, or those with an R on T phenomenon require therapeutic intervention. The primary means of treatment are the institution of lidocaine. Lidocaine is usually given as a 1 mg/kg dose followed by a continuous drip between 2-4 mg/min. When these arrhythmias occur one should check for hypoxemia, for an inappropriate potassium serum level, and for acid base imbalance. Problems with any of these should be corrected immediately. Ventricular premature contractions may also be controlled by ventricular overdrive pacing. For more long term control of persistent ventricular premature contractions, oral procainamide, quinidine or other drugs may be useful.

Ventricular Tachycardia

Ventricular tachycardia is a medical emergency. The immediate institution of lidocaine followed by countershock if the arrhythmia continues should occur. These patients will require careful and continual cardiac rhythm monitoring.

Ventricular Fibrillation

This disturbance requires immediate medical intervention. The treatment requires rapid asynchronous defibrillation. For patients who have had prolonged periods of ischemia associated with this rhythm disturbance, IV bicarbonate, epinephrine and defibrillation may be all required.

PULMONARY CARE

General Principles

All cardiac patients return to the CTU on a ventilator. Their parameters are:

1. FIO_2 of 100%
2. tidal volume between 10-15 cc/kg

3. rate of between 10-14 breaths/min
4. PEEP of between 2-8 mmHg

The respiratory adjustments are then made based upon:

1. a clinical assessment of the patient, and
2. the arterial blood gas analysis

Postop cardiac patients are usually extubated between 4-24 hours postop.

Ventilated Patient Requirements

All patients on ventilators require proper humidification of inspired gases. This is necessary to keep tracheobronchial secretions liquefied so that they can be removed.

All patients on ventilators require frequent percussion and vibration of their chest walls. This helps to eliminate thickened endobronchial and tracheal secretions.

All patients in their early postoperative course require frequent endotracheal suctioning to remove secretions accumulating in the tracheobronchial tree. We insist that our patients on ventilators undergo frequent turning and positioning so as to prevent hydrostatic pulmonary atelectatic changes.

Individuals who on clinical examination demonstrate bronchospastic airway findings should have added to their postoperative inhalation therapy treatment antibronchospastic agents such as bronchosol, aminophylline, etc.

INDICATIONS FOR EXTUBATION

All patients who are to be extubated on our service should meet the following criteria.

1. They should be alert with an adequate cough and gag reflex. In addition, they must be able to lift their head off the pillow.
2. They must have adequate cardiac, renal and nutritional status.
3. Their vital capacity must be between 10-15 cc/kg.
4. They must have a maximum inspiratory force less than -20 cm water.
5. Their alveolar arterial oxygen difference must be less than 350 mmHg.
6. Their pO_2 on an FIO_2 of 40% must be greater than 70 mmHg.
7. Their pCO_2 should be between 35-45 mmHg.
8. The pH if their arterial blood must be between 7.38 and 7.45.
9. Their base deficit should be normal.
10. They should have no life threatening dysrhythmias.
11. Their mediastinal chest tube drainage should be minimum.
12. Their tracheobronchial secretions should be manageable.
13. Their chest radiograph should be satisfactory.

POST-EXTUBATION CARE

Ensure that after removing the endotracheal tube the patient is breathing comfortably without any evidence on physical examination of pulmonary, cardiac or mental derangement.

Verify that the patient is doing well by obtaining an arterial blood gas 15-20 minutes after the tube has been removed.

Continue vigorous chest physiotherapy.

Prevent gastric dilatation. Note all cardiac surgical patients will have a nasogastric tube in place on return from the operating room to prevent this phenomenon. In addition, the nasogastric tube should be maintained on a long term basis in individuals who have demonstrated the inability to tolerate food or liquids by mouth. Furthermore, patients who have a weakened or absent gag reflex should have prolonged nasogastric tube suction to prevent aspiration pneumonitis.

All patients should be continued on incentive spirometry until they leave the hospital.

Once the endotracheal tube is removed, a chest radiograph should be obtained. Note that all postoperative cardiac patients will have a chest x-ray done immediately upon arrival to the Intensive Care Unit and then every morning for the first 3 postoperative days.

RENAL CARE
Low Urine Output
Low urine output (i.e., urine less than 1/2 cc/kg/hr) is considered a medical emergency in postop cardiac patients. Attention should be turned to the immediate diagnosis of the etiology of diminished urine output. Having diagnosed the problem, immediate therapy is indicated to prevent significant postoperative renal failure.

The Etiology of Low Urine Output/Renal Failure
Pre-Renal Factors

Under this grouping are problems with the delivery of blood to the kidney that results in a drop of urine output. This is described as a "pre-renal failure" type problem. Conditions which lead to this problem are hypovolemia, low cardiac output and cardiac tamponade.

Renal Factors

Under this condition are problems which result in direct injury to the kidney parenchyma and cause decreased urine output. Under these conditions we find the following:

1. Acute tubular necrosis which occurs secondary to prolonged low flow state, hemolysis with resultant hemoglobinuria and toxic injury to the kidney tubular system, and chemical injury to the kidney tubular system from agents such as gentamycin, angiographic dye, etc.
2. Embolic injury to the kidney caused by particulate emboli occurring during cardiopulmonary bypass or from clots or atherosclerotic debris released from an indwelling intra-aortic balloon pump.

Hepato-Renal Syndrome

This occurs in patients with profound liver disease. The exact mechanism for renal injury is not known.

Intrinsic Renal Pathology

Hypertensive nephropathy, nonsteroidal antiinflammatory nephropathy, diabetic Kimmelstiel-Wilson nephropathy, etc. predispose an individual patient to increased incidence of renal failure postoperatively.

Post-Renal Factors

This involves problems with the distal output. The most common cause for this condition is obstruction at the level of the bladder or bladder outlet. Benign prostatic hypertrophy is a common cause as is an obstruction of the Foley catheter due to clots or debris.

Complications of Renal Failure

Significant problems occur in the presence of postoperative renal failure. The recognition and treatment of these conditions is mandatory.

Hyperkalemia can develop rapidly in a patient with a low urine output. This problem predisposes individuals to significant ventricular arrhythmias and cardiac arrest. Treatment depends on the absolute serum potassium level. Levels in the 5.5-6 mEq/L range require immediate reduction in the serum potassium level. This can be obtained on a temporary basis by giving:

1. calcium gluconate 10 cc of a 10% solution IV push
2. giving 44 mEq sodium bicarbonate IV push
3. giving 1 amp 5% D5W plus 10 units of regular insulin IV push

When a more permanent reduction in the potassium level is required kayexelate 50 grams in 70% sorbitol solution can be given as a retention enema repeated on a Q1-2 hour basis. Each exchange will reduce the potassium level by approximately 200 mEq. If permanent reduction in the potassium level is required, or if the level becomes uncontrollable with the above modalities, either peritoneal dialysis or hemodialysis must ensue.

Pulmonary dysfunction frequently accompanies renal failure. This is caused by a combination of excessive fluid retention and increased susceptibility to pulmonary infections. The treatment of this problem, therefore, is control of excessive total body water in addition to antibiotic treatment of any existing pneumonias.

CNS disturbances frequently occur in the presence of ongoing renal failure. Increasing concentration of metabolic poisons produces somnolence and seizures in many patients. Therefore, treatment of the underlying problem by either hemodialysis or peritoneal dialysis can alleviate this situation.

Hematologic changes such as anemia and thrombocytopenia are produced by prolonged renal failure. Bleeding occurring secondary to thrombocytopenia may require infusion of platelets and/or red blood cells.

Acidosis of a renal origin always accompanies significant renal failure. Initial attempts to control this problem with infusion of bicarbonate may often stabilize the condition. However, should the acidosis become more profound, the institution of hemo- or peritoneal dialysis may be required.

Treatment of Renal Failure

Accurate assessment of the cause of decreased urine output, be it congestive heart failure, hypovolemia, cardiac tamponade, gentamycin nephrotoxicity or obstructed Foley catheter must ensue prior to instituting therapy. Once the predisposing condition is identified, treatment must be directed toward alleviating the cause.

10

Oliguria itself is an emergency. Attempt should be made to increase the urine output as early as possible for evidence exists showing that increasing the urine output of a failing kidney may significantly speed up the recovery of the organ.

Furosemide and ethacrynic acid have been shown to increase both water and solute excretion and have been documented to improve overall renal recovery by increasing the urine output.

Mannitol, by increasing renal cortical blood flow and enhancing sodium and water secretion, also has shown beneficial effects in an injured kidney.

Acidosis resulting from renal failure should be treated with sodium bicarbonate. In addition, renal injury by agents such as myoglobin or hemoglobin precipitation in the renal tubules has been shown to be markedly improved by alkalinization of the urine.

Medications which have either been shown to injure the kidney or to be excreted by the kidney must have their doses changed. It is clear that nephrotoxic drugs should be discontinued while drugs that are cleared by the kidney must have an altered dose regimen. Hyperkalemia must be treated, and potassium chloride infusions should be stopped.

INDICATIONS FOR DIALYSIS

The absolute indications for either hemo- or peritoneal dialysis in the postoperative cardiac patient are only three in number:

1. refractory acidosis
2. uncontrolled hyperkalemia
3. severe fluid overload

A lesser indication for dialysis may be considered significantly elevated blood urea nitrogen. In certain patients where the BUN raises to excessively high levels and evidence of CNS pathology may be occurring, an argument can be voiced for hemodialysis.

FLUID AND ELECTROLYTE MANAGEMENT

General Principles

All patients who have undergone cardiopulmonary bypass have significant shifts in their overall total body water. Cardiopulmonary bypass provides for significant dilution of the blood volume secondary to the crystalloid prime used to prime the pump. Cardiopulmonary bypass causes a significant increase in capillary permeability. These changes occur because of activation of the complement system, release of histamine from mast cells, release of serotonin from platelets, release of lysosomal enzymes from granulocytes and because of activation of the kinins. The overall result of these changes is marked increase in interstitial fluid. Following cardiopulmonary bypass patients can have an increase in the interstitial fluid volume of up to 150 cc/kg of body weight. One may safely assume that all patients who have undergone cardiopulmonary bypass have a significant interstitial fluid load which must be relieved in the first few postoperative days.

Fluid Management

The objectives of postoperative fluid management are:
1. to maintain an adequate intravascular volume
2. to reduce the excessive interstitial fluid overload
3. to prevent congestive heart failure and subsequent pulmonary edema

We are able to evaluate the postoperative fluid balance of an individual patient by:
1. obtaining accurate central venous pressure, pulmonary capillary wedge pressure, pulmonary diastolic pressure
2. assessing the adequacy of cardiac output
3. evaluating the patient input-output record
4. obtaining daily weight records and comparing them to preoperative weight
5. determining the postoperative hematocrit
6. determining the postoperative hourly urine output and specific gravity
7. clinical examination of the patient

Management is governed by the following principles:
1. Patients who have a low central venous pressure and pulmonary artery diastolic pressure in conjunction with a low cardiac output and a low hematocrit should have packed red blood cells infused to increase the intravascular volume.
2. Patients with a low CVP, a low pulmonary artery diastolic pressure and a low cardiac output who have a normal or high hematocrit should have one of two modes of therapy.

10

In the early postoperative period where significant interstitial fluid volume exists, 12.5-25 grams of mannitol is given IV to shift fluid from the interstitial to the intravascular space and subsequently to the urine. Infusion of 50 cc of 25% albumin accomplishes the same shift of fluid from the interstitial to the intravascular space; however, it does not provide an adequate diuresis. IV lasix 10-40 mg IV push may be given both in conjunction with mannitol and in conjunction with albumin to ensure the loss of the excessive interstitial fluid volume.

Patients who are further postop and who have demonstrated by examination of the I & O record and the daily weights that their interstitial fluid volume is normal, yet who still have a low filling pressure and evidence for intravascular fluid depletion, should be treated with intravascular volume replacement. It is our choice to use fluids that have their own oncotic pressure so that they remain in the intravascular space for prolonged periods of time. Our preference is to give 250 cc volumes of 5% albumin or equivalent volumes of Hespan.

In individuals in whom the CVP and pulmonary artery diastolic pressures are elevated, the cardiac output is down, and clinical evidence suggests congestive heart failure or pulmonary edema, efforts should be turned to reducing intravascular volume. Under these circumstances, furosemide or ethacrynic acid diuresis is the treatment of choice. In addition, digoxin for patients with significant histories of congestive heart failure may be utilized. Oxygen via mask or nasal prongs may be used to improve arterial oxygen saturation while morphine may be given to rapidly decrease preload and help bring the patient out of his congestive heart failure state.

In addition to the above one must be careful to evaluate the serum and urine glucose levels on patients known to be diabetic. Excessive glucose loss in the urine can provide for rapid intravascular depletion due to the osmotic diuresis. The situation should be quickly recognized and treated with insulin.

ELECTROLYTE MANAGEMENT

Postoperative electrolyte status should be evaluated by: (1) serum and urine electrolytes; (2) serum and urine osmolality determinations; (3) serum and urine creatinine and BUN; (4) by performing a good daily clinical examination. Therapy regulating disturbances in fluid electrolyte status should be as follows:

1. In conditions where there is evidence of excessive total body water with a drop in serum sodium, serum osmolality and increase in body weight, fluid restriction is indicated. These individuals should be put on fluid limits between 1,000 and 1,500 cc per day.
2. In cases where there is evidence of intravascular volume depletion with elevated serum sodium, serum osmolality, and elevated urine specific gravity and osmolality with a diminution in urine sodium, intravascular volume replacement as described previously is indicated.
3. Due to the obligatory diuresis that all of our patients undergo, significant urinary potassium loss may be anticipated. It is therefore mandatory that all postoperative cardiac surgical patients have frequent serum potassium determinations and IV bolus infusions of potassium chloride to maintain stable normal levels. It is our policy to give potassium as 10 mEq IV boluses in 50 cc D5W over an hour. Several of these boluses are usually given the first and second postoperative days.

POSTOPERATIVE BLEEDING

Evaluation and Management

The etiologies of postoperative hemorrhage are numerous:

1. Patients that are on antiplatelet agents or anticoagulants such as coumadin preoperatively tend to have problems with postoperative bleeding.
2. Individuals with known familial clotting defects such as von Willebrand's disease, hemophilia A or B, etc., clearly have problems with postoperative hemorrhage.
3. Individuals who have undergone reoperation, because of the extensive dissection required and the usually longer bypass run, bleed much more in the postoperative state.
4. Patients that have prolonged complicated operations have an increased incidence of postoperative bleeding. This is due to the increased destruction of platelets and clotting factors by the cardiopulmonary pump apparatus.

5. Technical factors, such as a leak in a suture line or poorly tied ligature, can produce brisk postoperative hemorrhage.

6. Individuals who suffer transfusion reactions can spontaneously hemorrhage in the postoperative period.

7. Hypertension is an enemy of the cardiac surgeon. Frequently, patients who are not bleeding develop an episode of significant hypertension and start bleeding via the mediastinal tubes. It is obvious that clots have been forced off injured vessels or hemorrhage has been produced through suture lines due to the increased systolic blood pressure.

Medical therapy to control postoperative hemorrhage should be directed toward the specific problem.

If a coagulopathy is suspected one should obtain PT and PTT to evaluate the clotting cascade of both the intrinsic and extrinsic systems. Fresh frozen plasma should be given to correct clotting defects.

If platelet function is felt to be the problem, one should first obtain a platelet count. If this is low, infusion of 10 units of platelet concentrate should be given. If, however, the platelet count is normal, but one suspects platelet function disorder (e.g., from preop aspirin use), then platelet concentrate should also be given. It is difficult, if not impossible, to obtain platelet function tests on an urgent basis.

Heparin rebound may be the problem. This occurs because of three basic problems:

1. Heparin is bound to fatty tissues which later release it after protamine reversal.

2. The heparin-protamine complex is unstable with the half life of protamine being significantly less than heparin, therefore allowing heparin to once again circulate free in the blood stream.

3. Peripheral tissues which have been isolated from the main flow of blood due to vasoconstriction secondary to hypothermia may later become perfused releasing the heparin which is sequestered there.

Irrespective of the cause of hemorrhage, one may evaluate it by performing an activated clotting time (ACT). If the test is positive, heparin rebound may be treated with IV boluses of 50 mg of protamine until the ACT is corrected.

Fibrinolysis may be suspected in patients who have had prolonged cardiopulmonary bypass. When one suspects this situation, a serum fibrinogen level may be obtained. In addition, blood drawn from the patient and allowed to clot will show, if followed for a long enough period of time, dissolution of the clot. Under these circumstances, the treatment of primary fibrinolysis is IV administration of amicar. This drug is given as 5 grams IV push followed by 1 gram per hour until the bleeding stops.

With prolonged episodes of bleeding and clotting, depletion of the primary source of serum fibrinogen occurs. Under these circumstances, the serum fibrinogen level will be extremely low, and the treatment is IV infusion of 10 units of cryoprecipitate. Note: cryoprecipitate is the product of specially prepared serum which is used to treat hemophilia A. Though the concentrations of Factor VIII are high in this product, the concentration of fibrinogen are also high and approximately 15 times that of fresh frozen plasma.

Disseminated intravascular coagulopathy (DIC) may occur under various conditions such as sepsis. When this condition is suspected, serum fibrinogen level will be low. The fibrin degradation products will be high, the protamine sulphate test will be positive and the patient's platelet count will be low. The therapy is the IV infusion of heparin to stop the ongoing intravascular coagulation process. However, in the postoperative cardiac patient, this procedure may not be safe.

INDICATIONS FOR REOPERATION

The indications for reoperation are dependent upon the mediastinal chest tube drainage. We hold as our criteria three absolute indications:
1. mediastinal chest tube output of between 300-500 cc per hour for the first hour
2. 200-300 cc per hour for the second hour
3. greater than 100 cc per hour for 6-8 hours

Though the absolute indications for reoperation are listed above, the treatment of each patient must be individualized. Consideration must be given to the operative procedure that the individual patient has undergone. Individuals who have had prolonged complicated operations which may or may not have involved reoperation, where at the end of the case significant effort has been put forth to control bleeding, yet the patient is still bleeding in the postoperative course, should be given a much longer period of time with higher volumes of mediastinal output before returning to the operating room. Frequently, these individuals can be brought under control with intense medical management.

Evaluation of the chest radiograph and hemodynamics play a significant role in determining reoperation. Individuals who may have been bleeding and then suddenly stop, while their chest radiograph shows a widening of the mediastinum with CVP and pulmonary diastolic pressures showing significant equalization and elevation, suggest cardiac tamponade. Under these circumstances, an urgent trip to the operating room for mediastinal evacuation and control of hemorrhage is mandatory.

The absolute trend in bleeding also has a significant part to play. Patients bleeding dramatically who then seem to slow over a period of time may be given more time to see if medical management can effectively control the bleeding.

PREPARATION FOR REOPERATION

The junior resident should have instructions either from the chief resident or the attending staff to take the patient back to the operating room. The on-call resident should ensure that a slip has been turned in to the operating room so that they know the patient is coming. The patient's family should be informed that the patient is being returned to surgery and an informed consent should be obtained if possible. The house officer should notify all outside attending staff including the patient's private cardiologist that the return is occurring. The resident on call should alert the anesthesia department and again check with the operating room to see that they will be ready. The resident on call should be sure that adequate blood and blood products are available.

SUGGESTED READING

1. Baumgartner WA, Owens SG, Cameron DE et al, eds. The Johns Hopkins Manual of Cardiac Surgical Care. St. Louis: Mosby Year Book, Inc., 1994.
2. Blitt CD. Monitoring in Anesthesia and Critical Care Medicine. New York: Churchill Livingstone, 1985.
3. Kotler MN, Alfieri AD, eds. Cardiac and Noncardiac Complications of Open Heart Surgery: Prevention, Diagnosis, and Treatment. Mount Kisco, NY: Futura, 1992.
4. Daily EK, Schroeder JS. Techniques in Bedside Monitoring. St. Louis: Mosby, 1989.

10

General Thoracic Anatomy

Fritz J. Baumgartner

SURFACE ANATOMY OF THE LUNGS

The right lung is composed of three lobes—upper, middle and lower, and is the larger of the lungs. The left is made up of only two lobes, upper and lower, and is smaller because of impingement of the heart on the lung. Two fissures are present on the right—the major fissure and the minor fissure in the oblique and horizontal planes, respectively (Fig. 11.1a).

Variations in the fissures occur and often there is a failure of development of part or all the fissure. This is frequently seen as a less complete fusion between the middle lobe and upper lobe on the right side. Accessory fissures may occur and one example of this is the azygous lobe, an accessory lobe in the right upper lobe formed from a fissure developed from an aberrant loop of the azygous vein.

The anatomic pattern seen on a PA and lateral chest film is characteristic (Fig. 11.1b).

BRONCHOPULMONARY SEGMENTS

The general pattern is a total of 18 segments—10 in the right lung and 8 in the left (Fig. 11.2). The main difference is that on the left side, the apical-posterior segment of the left upper lobe is considered as one segment, whereas apical and posterior segments of the right upper are considered as two separate segments. Also on the left side, there is no medial segment of the left lower lobe because of impingement of the heart on the left lower lobe.

Another difference is that on the left side, there is no middle lobe, rather a lingular portion which consists of a superior and inferior segment. This is unlike the right side where the middle lobe consists of a medial and lateral segment.

THE BRONCHIAL TREE

The trachea bifurcates at about the level of the fourth thoracic vertebra into the right and left mainstem bronchi. The tracheal bifurcation at the carina is considered a major anatomical landmark where several anatomic events occur. These include the origin of the great vessels coming off the aorta and the entry of the

11

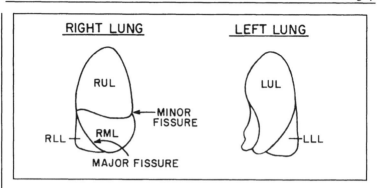

Fig. 11.1a. Surface anatomy of the lungs, viewed anteriorly.

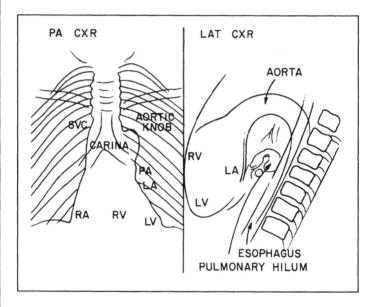

Fig. 11.1b. PA and lateral x-rays of a normal chest and line diagrams of the anatomy (SVC-superior vena cava, RA= right atrium, RV= right ventricle, LV= left ventricle, LA= left atrium, PA= pulmonary artery).

azygos vein into the superior vena cava at the level of the angle of Lewis. These occur at the level of the T4 vertebral body.

The trachea bifurcates into a right mainstem bronchus which arises in a more direct line with the trachea compared to the left bronchus which arises at a sharper angle to the trachea. Hence, aspiration pneumonia usually occurs in the right lung because the angle is more favorable for this event to occur. The right mainstem bronchus is much shorter than the left, and the origin of the right upper lobe

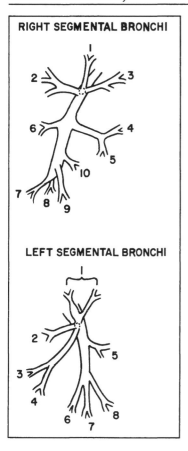

RIGHT SEGMENTAL BRONCHI

LEFT SEGMENTAL BRONCHI

Fig. 11.2. Segmental anatomy of the lung. (RUL: 1-Apical; 2-Post; 3-Ant); (RML: 4-Lat; 5-Med.); (RLL: 6-Sup; 7-Lat Basal; 8-Post Basal; 9-Ant Basal; 10-Med Basal); (LUL: 1-Apical-Post; 2-Ant; (3-Sup; 4-Inf Lingula)); (LLL: 5-Sup; 6-Post Basal; 7-Ant Basal; 8-Lat Basal)

11

orifice is only about 1.2 cm away from the carina. Proceeding distally from the takeoff of the right upper lobe bronchus is the bronchus intermedius, i.e., that portion of the bronchus between the right upper lobe takeoff and the middle lobe takeoff. The right pulmonary artery crosses over the mainstem bronchus, hence the term "eparterial bronchus" to designate the right mainstem bronchus which is superior and posterior to the right pulmonary artery. At the level of the middle lobe bronchus is the superior segmental bronchus of the lower lobe. The middle lobe bronchus arises from the anterior surface of the bronchus intermedius, whereas the superior segmental bronchus of the lower lobe arises from the posterior wall of the bronchus intermedius only slightly distal to the middle lobe bronchus and indeed may appear to be at the same level. Thus, care must be taken in performing a lower lobectomy or a middle lobectomy so as not to injure that bronchus leading to a normal lobe. The left mainstem bronchus is longer than the right and branches about 5 cm distal to the carina. The left mainstem is "hyparterial", i.e., inferior and anterior to the left pulmonary artery. The upper lobe bronchus divides into an upper lobe proper branch and a more inferior branch; this is the lingula branch. About half a centimeter distal to the left upper lobe orifice, the lower lobe bronchus gives off its first branch—the superior segmental bronchus. The lower lobe bronchus then continues down to the branch into the basal segmental bronchi. Bronchoscopic segmental anatomy is shown in Figure 11.3.

PULMONARY ARTERIAL SYSTEM

The mainstem pulmonary artery trunk arises to the left and posterior to the aorta and then branches into the right and left pulmonary artery. The right pulmonary artery passes directly posterior to the aorta. The right pulmonary artery is longer than the left. No one pattern for either the right or left pulmonary artery is "standard"; however, a relatively uniform distribution occurs for each. The right pulmonary artery is seen in Figure 11.4. The first branch is the truncus anterior branch which divides into apical anterior and posterior segmental arteries. The posterior artery from the truncus anterior supplies only a part of the posterior segment since there is an additional posterior ascending artery arising further down from the pulmonary artery and is found in 90% of people. This is found within the fissure unlike the truncus anterior branch which is found outside the fissure. The posterior ascending artery is important, however, since it must be identified and ligated when performing a right upper lobectomy to prevent major hemorrhage from injuring this unsuspected branch.

The middle lobe artery arises from the right pulmonary artery at the same level as the posterior ascending artery and also at the same level as the superior segmental artery and therefore care must be taken when taking a lobar pulmonary artery to prevent damage to another lobe. The left pulmonary artery is seen in Figure 11.5.

The left pulmonary artery is posterior and superior to the left mainstem bronchus, i.e., the left mainstem bronchus is hyparterial or below and anterior to the left PA, unlike the right side where it is eparterial.

The left pulmonary artery branches into an upper lobe pulmonary artery with an apical artery, an anterior artery and coming off distally, separate posterior arteries. The superior segmental artery to the left lower lobe actually comes off proximal to the takeoff of the lingula artery. This is the most common variant, although a third of the time the superior segmental artery may come off distal to the lingula artery. The left pulmonary artery then continues to the basal artery.

PULMONARY VENOUS SYSTEM

The venous drainage of the lung is more variable than the arterial pattern. There are two major venous trunks from both lungs—the superior and inferior pulmonary veins.

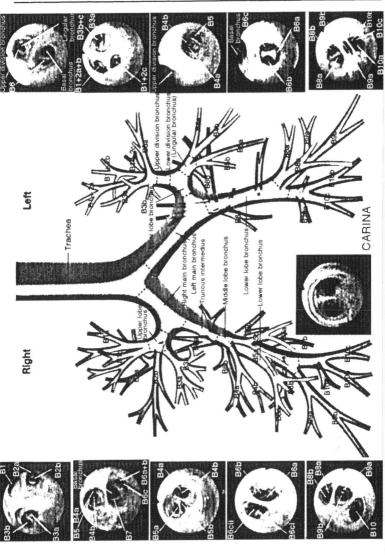

Fig. 11.3. Bronchoscopic anatomy of the lung. Right upper lobe: B1-apical segment; B2-post segment; B3-ant segment. Middle lobe: B4-lat segment; B5-med segment. Lower lobe: B6-sup segment; B7-med segment; B8-ant segment; B9-lat segment; B10-post segment. Left upper lobe: B1 & 2-apical & post seg; B3-ant segment. Lingula: B4-sup segment; B5 inf segment. Lower lobe: B6-sup segment; B8-ant segment; B9-lat segment; B10-post segment. From: Olympus Chart. Kenkichi Oho, M.D., PhD; Tokyo Medical College.

11

Fig. 11.4. Anatomy of the right pulmonary artery.

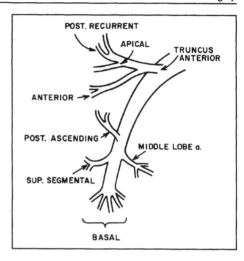

Fig. 11.5. Anatomy of the left pulmonary artery.

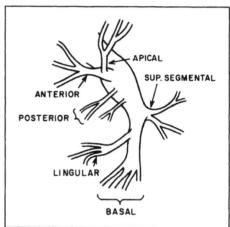

On the right side, the superior pulmonary vein lies anterior and inferior to the pulmonary artery. The upper lobe pulmonary vein and middle lobe pulmonary vein drain into the right superior pulmonary vein. On occasion, the middle lobe pulmonary vein may enter the pericardium and drain into the atrium as a separate vessel and even more rarely is a tributary of the inferior pulmonary vein. The inferior pulmonary vein is inferior and posterior to the superior vein and drains the lower lobe. The inferior pulmonary vein is at the superiormost portion of the inferior pulmonary ligament; this is an anatomical point of major surgical significance.

On the left side, there is again a left superior pulmonary vein and a left inferior pulmonary vein. As on the right, the superior pulmonary vein is anterior and inferior to the left pulmonary artery. The left inferior pulmonary vein, as on the right, is located inferiorly and posteriorly to the superior vein at the upper most portion of the inferior pulmonary ligament.

BRONCHIAL ARTERIES AND VEINS

The bronchial arterial system arises from the systemic circulation directly off the aorta and its branches. It empties primarily into the pulmonary veins and a lesser bronchial vein system that enters the azygos venous system on the right and hemiazygos on the left. The origins of the bronchial arteries are variable and come from the aorta, intercostal arteries and occasionally subclavian and innominante arteries. In most people, there is one bronchial artery on the right and two on the left, although there may be one on each side or two on each side. The major source on the right is from the first or sometimes the second intercostal artery. On the left, the bronchial arteries usually take their origin directly from the aorta.

LYMPHATIC SYSTEM

The lymphatic system of the lung includes intrapulmonary lymph nodes (N1) and mediastinal lymph nodes (N2). The intrapulmonary nodes are within the visceral pleural envelope. These are shown in Figure 11.6. The mediastinal lymph nodes are also shown in Figure 11.6. They can be classified numerically to include node stations 1, 2, 3, 4, 5, 6, 7, 8 and 9. It should be noted that level 10 nodes may be considered either intrapulmonary or mediastinal, depending on the level that they are sampled at. They are more likely to be intrapulmonary on the left side because the left mainstem bronchus is longer than that on the right.

With regard to the N2 nodes, these are mediastinal nodes and can be also classified into the following four categories:
1. Paratracheal nodes: These include nodes around the level of the trachea and include level 1, 2 and 3 nodes.
2. Tracheobronchial nodes include node stations 4, 10 and 7. The level 7 node is an important station since it is in the subcarinal location. More will be described about this later.
3. Posterior mediastinal nodes: These include levels 8 and 9 which are paraesophageal and inferior pulmonary ligament nodes.
4. Anterior mediastinal nodes: These include the aortopulmonary and para-aortic nodes (levels 5 and 6).

As mentioned, level 7 or subcarinal nodes represents an important nodal station. This can be sampled by mediastinoscopy as can the other paratracheal nodes and tracheobronchial nodes, and this will be described later. However, only the anterior portion of the level 7 subcarinal node can be biopsied by mediastinoscopy and, in this sense, it represents only the tip of the iceberg.

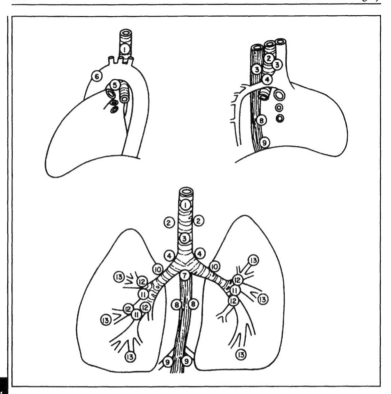

Fig. 11.6. Lymph node map of mediastinal (N2) and intrapulmonary (N1) nodes. 1, 2, 3. N2 Nodes (Mediastinal): 1= high pre/retrotracheal; 2= high paratracheal; 3= low pre/retrotracheal; 4= low paratracheal; 5= aortopulmaonary; 6= paraaortic; 7= subcranial; 8= paraesophageal; 9= pulmonary ligament; 10= hilar. N1 Nodes (Intrapulmonary): 10= hilar; 11= interlobar; 12= lobar; 13= segmental. Another classification is anterior mediastinal (5, 6); posterior mediastinal (8, 9); tracheobronchial (7, 4, 10); and paratracheal (1, 2, 3).

The level N2 nodes, if involved, means mediastinal involvement by lung carcinoma, and signifies unresectibilty. Conversely, N1 nodes are intrapulmonary nodes and although involvement of these nodes by carcinoma implies a worse prognosis and decreased life expectancy, it does not preclude resectability. The lymphatic drainage of the right lung is ipsilateral, except for an occasional instance in which drainage to the superior mediastinum is bilateral. Drainage from the left lung to the superior mediastinum is as frequently ipsilateral as it is contralateral.

Another important concept is the "lymphatic sump of Borrie" which represents lymph nodes on each side that receive lymphatic drainage from all lobes of the corresponding lung (Fig. 11.7). The significance is that nodes in this region can be involved with lung carcinoma from any lobe of the lung on that particular side.

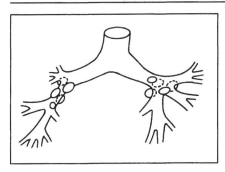

Fig. 11.7. Sump of Borrie.

SUGGESTED READING

1. Brock RC. The anatomy of the bronchial tree. 2nd ed. London: Oxford University Press, 1954.
2. Corey RAS, Valentine EJ. Varying patterns of the lobar branches of the pulmonary artery. Thorax 1959; 14:267.
3. Kent EM, Blades B. The surgical anatomy of the pulmonary lobes. J Thoracic Surg 1942; 12:18.
4. Shields TW. Surgical anatomy of the lungs. In: Shields, ed. General Thoracic Surgery. 5th ed. Philadelphia: Lea and Febinger, 2000:63-75.

11

Lung Cancer

Fritz J. Baumgartner

GENERAL

Those patients who have an endobronchial mass on endoscopy may have a benign lesion or a malignant lesion. Of the malignant lesions, this can be low grade malignant lesions, one of the more common malignant lesions, or a rare malignant lesion. Benign lesions include such things a plasma cell granuloma (inflammatory pseudotumor). This is a benign tumor that is a post-inflammatory reaction and not a true neoplasia. Other lesions include granular cell myoblastomas or granular cell tumors. These originate from Schwann cells and are usually found in larger bronchial origins although they may occur as a peripheral lesion. Leiomyoma, lipoma and fibroma are other benign mesenchymal tumors. The more common malignant tumors include large cell carcinoma, adenocarcinoma, small cell and squamous cell carcinoma. Bronchoalveolar carcinoma can be considered a subset of adenocarcinoma.

The low grade malignant lesions include carcinoid tumor. This is a cancer, albeit a low grade cancer; it is not a benign lesion. There is a continuum histologically of carcinoid up to the level of small cell carcinoma. This series is actually a family of tumors called Kulchitsky cell carcinomas. Other low grade malignant lesions include adenoid cystic carcinoma. This is characterized by submucosal spread and perineural invasion such that even though a lesion may be seen endobronchially, usually in a centrally located bronchus, this may have spread submucosally. Another histologic type of low grade malignancy is mucoepidermoid carcinoma.

NONSMALL CELL LUNG CANCER

Carcinoma of the lung is the most common cause of death due to cancer in both men and women. Lung cancer can be divided into two fundamental types—small cell lung cancer (which is synonymous with oat cell lung cancer) and nonsmall cell lung carcinoma. The differentiation is clinically important as small cell lung carcinoma has a much lower survival rate than nonsmall cell carcinoma.

The histologic subtypes of nonsmall cell lung carcinoma include adenocarcinoma, large cell carcinoma and squamous carcinoma. There is another histologic subtype called bronchoalveolar carcinoma which is most likely a subtype of

Table 12.1. Staging of NSCLC

Stage		Survival	
I ——————— a T1, N0		85%	
———— b T2, N0			Chance for surgical management
II ——————— a T1, N1		60%	
———— b T2, N1 or T3 N0			
III —— a T3, N1, or Any N2		20%	
—— b Any T4 or Any N3			
IV M		<10%	Medical management

T1 <3 cm	N1 intrapulmonary nodes
T2 >3 cm or invading visceral pleura or obstructing atelectasis/pneumonitis	N2 ipsilateral mediastinal nodes
T3 invading nonessential mediastinal structures or within 2 cm of carina	N3 contralateral nodes; ipsilateral or contralateral supraclavicular nodes.
T4 invading essential mediastinal structures or malignant pleural effusion	

adenocarcinoma, but has a better prognosis overall. Adenocarcinoma of the lung is the most frequent histologic type, responsible for 50% of lung cancers. Squamous cell is the next most frequent cancer accounting for 30%, and small cell lung cancer represents 15%. Large cell cancer makes up less than 5%.

The rare primary malignant tumors include sarcomas (which may be parenchymal or endobronchial), or lymphoma (either Hodgkin's or nonHodgkin's). Only 0.3% of lymphomas actually originate in the lung.

NONSMALL CELL CARCINOMA STAGING

The stages are depicted in Table 12.1. These are stage I, II, III and IV. The 5 year disease free survival for stage I is in the range of about 85%; that for stage II is 60%; that for all types of stage III is about 30%; and that for stage IV less than 10%. The stages are identified by various classifications within the TNM system. T1 is a lesion of less than 3 cm without invasion of the visceral pleura. T2 is a lesion greater than 3 cm or any sized tumor that invades the visceral pleura or has associated atelectasis or obstructive pneumonitis extending to the hilar region. T3 is a tumor of any size with direct extension into nonessential chest structures. This includes invasion into the chest wall (including superior sulcus tumors), diaphragm, mediastinal pleura, and pericardium without heart involvement. Also a

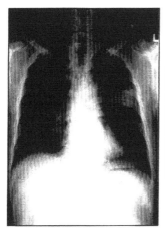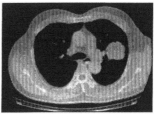

Fig. 12.1. CXR (a-left) and CT scan (b-right) of a patient with a large T2 lung tumor. The staging remains stage I because there is no other apparent invasion or nodal involvement. Because of the size of the lesion, it is in stage Ib.

tumor in the mainstem bronchus within 2 cm of the carina but not involving the carina is a T3 lesion. A T4 lesion denotes invasion of essential mediastinal structures. These include heart, great vessels, vertebral body, esophagus, trachea or carina. Also the presence of a malignant pleural effusion is a T4 lesion.

The nodal classification is as follows: N0 means no nodal involvement. N1 implies bronchopulmonary nodes according to the old classification system. These include any nodes within the visceral pleural envelope. N2 nodes involve tracheobronchial nodes according to the old classification system, i.e., mediastinal nodes outside of the visceral pleural envelope. N3 nodes are contralateral mediastinal nodes, contralateral hilar nodes, or ipsilateral or contralateral scalene or supraclavicular lymph nodes. The presence of metastases indicates M1 disease.

The following is a redefined staging classification as instituted by Mountain in 1997. Stage I is a T1 or T2 lesion without nodal involvement (Fig. 12.1 a-b) or a T2 lesion with N1 pulmonary node involvement. Additionally, a T3 lesion without nodal involvement is now considered a stage II lesion. This is the most important change of Mountain's new classification compared to pre-1997 classification schemes. Previously T3, N0 lesions were classified as stage III leasions, but with appropriate therapy it became clear that these T3, N0 lesions had a far better prognosis than their stage III counterparts, prompting the change to stage II.

Stage III represents a far more grave category. Stage IIIa is T3 disease with intrapulmonary (N1) lymph node involvement or any mediastinal (N2) involvement. Surgery may still be offered in T3, N1 disease, e.g., superior sulcus tumors with N1 nodes not extending to the mediastinum. Generally, surgery is not offered for lung carcinoma with N2 mediastinal involvement. Stage IIIb implies unresectability and includes either a T4 or N3 lesion. A stage IV lesion is any lesion with metastases.

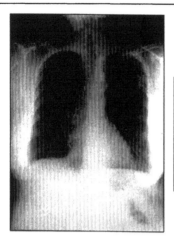
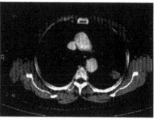

Fig. 12.2. CXR (a-left) and CT scan (b-right) of a patient with a T3 lung tumor with chest wall involvement. If subsequent lobectomy revealed no intrapulmonary nodes (N0), it would be stage IIb; if there were intrapulmonary nodes (N1) it would be stage IIIa.

Generally, stages I and II are managed primarily by surgical resection. Stage IIIa lesions may or may not be managed surgically. Examples of lesions that can be resected surgically include superior sulcus tumors or other peripheral lung tumors with chest wall involvement (Fig. 12.2a-b). However, N2 lesions involving the mediastinal nodes are generally considered unresectable, yet these are still stage IIIa lesions. Stage IIIb lesions and stage IV lesions are generally considered unresectable.

What about recurrent laryngeal nerve involvement or phrenic nerve involvement? Does this imply an unresectable tumor? Frequently, lesions involving the recurrent laryngeal nerve are in the aortopulmonary window and involve the mediastinal nodes in positions 5 and 6, near the recurrent nerve on the left side. Some feel that this may be resectable but most surgeons feel that this is an unresectable lesion. With regard to phrenic nerve involvement, this may be resectable if only the pericardium is involved without actual extension onto the heart. A malignant pleural effusion implies an unresectable T4 lesion which is a stage IIIb.

SMALL CELL LUNG CANCER

Small cell lung cancer is considered by many to be a systemic disease since at initial presentation nearly two-thirds of patients have disease outside the thorax involving at least one other organ. It is felt by some that the diagnosis of small cell carcinoma in the lung, in and of itself, is not resectable because of presumed systemic metastases. Regardless of the ability to document all sites of micrometastases, chemotherapy is considered the first line of treatment for small cell carcinoma.

Because of this, the TNM classification is occasionally not used for small cell carcinoma; it is presumed that all such cases have micrometastases systemically. In any event the TNM system should be employed for small cell carcinoma since most thoracic surgeons now feel that there are specific indications for resecting small cell lung carcinoma. These indications, for example, may include a patient with a stage I peripheral small cell carcinoma without mediastinal involvement. Resection may result in cure. Some groups have reported a 5 year survival rate of almost one-third following surgical management of small cell lung cancer.

Surgery in small cell lung cancer is limited to about 10% of all patients. Surgical management may play a role in the treatment of small cell lung carcinoma but must be done in conjunction with chemotherapy, either preoperative or postoperative, or both, with or without radiation therapy.

Currently, resection of stage I and II small cell lung cancer with pre- or postoperative chemotherapy is the recommended treatment with additional mediastinal radiation therapy of N1 positive disease. No resection is indicated if N2 disease, i.e., mediastinal disease, is involved. Also, prophylactic cranial radiation is performed in cases of small cell lung cancer because of the high risk of brain metastases.

Chemotherapy regimens generally used for small cell lung cancer include cyclophosphamide, doxorubicin and vincristine. Without treatment, small cell lung cancers are rapidly fatal with the mean duration of survival less than 3 months.

PARANEOPLASTIC SYNDROMES

The extrathoracic paraneoplastic syndromes are neuroendocrine phenomena associated with lung carcinoma. These include the following:

1. Cushing's syndrome. Patients with small cell carcinoma can produce a large amount of ACTH. Dexamethasone fails to suppress the levels of ACTH and hence Cushing's syndrome secondary to lung carcinoma is distinguishable from a pituitary tumor producing ACTH which is suppressed by dexamethasone. The signs and symptoms of the syndrome may be suppressed by appropriate chemotherapy for small cell cancer.

2. Inappropriate ADH production. The symptoms are water intoxication with nausea and vomiting and worsening neurologic complications secondary to hyponatremia. This occurs most often in patients with small cell lung cancer.

3. Hypercalcemia caused by PTH-like polypeptide has most often been associated with squamous cell carcinoma. Hypercalcemia usually represents bony metastases; however, it may be caused by secretion of the PTH like hormone.

4. Myasthenia-like syndrome is due to a defect in neuromuscular conduction similar to that in myasthenia gravis but is found in patients with bronchogenic carcinoma. It is due to antibodies against the acetylcho-

line receptor produced by the tumor. This myasthenia-like syndrome is called the Eaton-Lambert syndrome.

5. Carcinoid syndrome. This has been reported in a few patients with oat cell carcinoma secondary to amine secretion by the tumor.

In addition to the above mentioned myasthenia-like Eaton-Lambert syndrome, other neuromuscular manifestations include polymyositis, peripheral neuropathy, encephalomyelopathy and cerebellar degeneration. Other manifestations include skeletal manifestations with generalized hypertrophic pulmonary osteoarthropathy with periosteal proliferation and new bone formation. The incidence has been reported to be between 2 and 12%. This phenomenon is not found in patients with small cell tumors, although its incidence in patients with the other three major cell types was equally distributed.

Dermatologic manifestations include acanthosis nigricans which may be associated with adenocarcinoma. Other manifestations include thrombophlebitis and may be considered part of the generalized Trousseau syndrome of hypercoagulable state associated with neoplasia.

THE APPROACH TO THE PATIENT WITH A LUNG MASS

HISTORY AND PHYSICAL EXAM

The patient who presents with the chest x-ray finding of lung mass must be evaluated to quickly and accurately determine the etiology of the lesion and establish a treatment plan. The history and physical examination are of prime importance. The patient is asked about symptoms of cough, sputum production, chest pain, shortness of breath and weight loss. Hemoptysis is an important finding. A previous history of tuberculosis should be elicited. Physical examination should include cardiopulmonary examination and a search for nodes in the axilla, supraclavicular regions and neck, as well as a general physical examination. Given a negative general history and physical examination without signs of another primary lesion, the chance of a solitary lung lesion being a metastasis from another organ is less than 5% and it is almost certainly a new primary in the lung.

The presence of tuberculosis should never be underestimated since it comes in many disguises. As with appendicitis for the abdomen, tuberculosis as a rule should never be less than second on a list of differential diagnoses for lung pathology. On a broader perspective, lung lesions in general can be categorized into tumors, infectious etiologies, trauma, congenital lesions or idiopathic problems. This will be described in more detail in relation to the solitary pulmonary nodule.

Inspection of the pulmonary mass on the chest x-ray should be carefully done. There are some clues that could give insight into the pathology of the lesion. Generally, tuberculosis is found in the apices of the lung although it can be found in any location in the lung. Cavitary lesions associated with carcinoma are generally squamous cell carcinomas. Squamous cell carcinomas and small cell carcinomas are usually, but not universally, centrally located whereas large cell tumors and adenocarcinomas are in general more peripherally located. The presence of pop-corn calcification, lamellar calcification or target calcification in general implies benign disease such as a granuloma, although carcinoma cannot be excluded.

A general laboratory work-up of the patient with a lung mass includes CBC and electrolytes. The fundamental work-up of lung mass then includes sputum cytology, CT scan of the chest and flexible fiberoptic bronchoscopy. These three tests should be included in the preoperative evaluation of all lung masses. CT scan is useful for a number of reasons, not only to further define the location and ex-tent of the lung lesion, but also to identify any synchronous lesions that may have been missed on the chest x-ray. Most importantly the CT evaluates for mediasti-nal adenopathy, the presence of which may require investigation. The use of CT scanning as a modality for staging of lung carcinoma is controversial. In general, in the United States the standard of care for staging of lung carcinoma is by CT scan. Mediastinal nodes greater than 1.0 cm require further evaluation. Node en-largement may represent either reactive inflammation or carcinoma. Further as-sessment of these nodes is done either by mediastinoscopy, anterior mediastino-tomy (Chamberlain procedure), or by video-thoracoscopic methods. Conversely, nodes less than 1.0 cm are generally considered negative; however both the sensi-tivity and specificity of CT scanning in staging of lung cancer is 85%. In other words, the chance of an enlarged node on CT scanning indeed being positive for carcinoma is only 85%, whereas the chance of a node less than 1.0 cm being nega-tive for carcinoma is again also 85%. In other countries such as Canada, mediasti-nal lymph node sampling by mediastinoscopy or mediastinotomy are considered routine.

The diagnosis of the lesion may be achieved by sputum cytology or by bron-choscopy with biopsy of the lesion or bronchial brushings, and/or bronchial washings. If the lesion still cannot be diagnosed, then several alternatives are avail-able. One is to perform a fine needle aspiration of the lung lesion. The problems with this technique is that there is a 10% chance of a pneumothorax and that a malignant lesion may result in a negative or nondiagnostic needle aspiration. This occurs approximately 15% of the time, and thus although a positive fine needle aspiration may be helpful, a negative fine needle aspiration may be harmful if it is incorrect. Therefore some advocate going straight to thoracotomy when the above work-up has not determined the etiology of the lesion, rather than performing fine needle aspiration. The one lesion where a fine needle aspiration is extremely helpful is in the case of a superior sulcus tumor (i.e., Pancoast's tumor). Manage-ment of this tumor involves preoperative radiation therapy followed by surgical resection of the chest wall en bloc with the tumor. Thus, positive aspiration cytol-ogy will prevent a diagnostic thoracotomy prior to radiotherapy. Other instances where fine needle aspiration may be very helpful is where there are suspicions for small cell carcinoma, lymphoma, or infectious etiologies (e.g., tuberclosis).

After all these preoperative diagnostic and staging procedures have been done, surgical resection may be considered. The primary questions to be answered are:

1. Is the patient a suitable candidate for thoracotomy and major pulmo-nary resection?

2. Is the lesion itself surgically resectable in terms of the local lesion itself and in terms of mediastinal node involvement?

Whether the patient himself is a suitable candidate for resection depends on the history and physical examination, arterial blood gas and pulmonary function tests, and evaluation of the patient's overall condition. If he is too cachectic and malnourished to undergo major thoracotomy, then this is an important consideration. Are there other co-morbid conditions such as synchronous cancer, coronary artery disease or peripheral vascular disease, or other problems that may make pulmonary resection dangerous?

The arterial blood gas is of prime importance in determining if patient is a candidate for surgery. The $PaCO_2$ is the most important factor in this determination. Generally, if the resting pCO_2 is greater than 45 mmHg operation is precluded because by definition, the patient is not ventilating adequately and will not have enough pulmonary reserve for a major lung resection. Another simple test which is a good assessment of pulmonary reserve is to take the patient up a flight of stairs and observe how the patient is able to tolerate this. If this is well tolerated, he generally has enough reserve to tolerate a major lung resection.

Pulmonary function tests are invaluable in assessing a patient for lung resection. The FEV-1 is the most important determinant. It is generally thought that a predicted postoperative FEV-1 of less than 0.8 liters precludes resection. This rule may be tempered by various circumstances. For example, a sedentary, old lady may do well with an FEV-1 of 0.8, whereas a younger man, otherwise active and healthy, may do poorly if his predicted postop FEV-1 is less than 1 liter. Although clinical judgment is important, the value of 0.8 liters is generally considered a boundary which should not be crossed. There are various ways to determine the postoperative FEV-1. For example, if a patient has a preoperative FEV-1 of 2.0 liters, a pneumonectomy will presumably reduce his FEV-1 by half with the postop FEV-1 being 1.0 liters. Using these determinants and knowledge of the segmental anatomy of the lungs, one will be able to determine what the resultant FEV-1 will be after resection of a lobe or segment.

To determine this, it is important to understand the segmental anatomy of the lungs. On the right side, there are three segments in the upper lobe, two segments in the middle lobe and five segments in the lower lobe. This makes a total of ten segments in the right lung. On the left side, there are four segments in the upper lobe (two segments in the upper lobe proper and two segments in the lingula), and there are four segments in the lower lobe making a total of eight segments. Overall, approximately 55% of the total lung function is contributed from the right lung and 45% of the total lung function is contributed from the left lung. One can thus assess whether to do a resection based on postoperative function predicted by the number of segments resected. For example, if one does a right upper lobectomy in a patient with an FEV-1 of 1.1 what will be the predicted FEV-1? In a right upper lobectomy, three segments are removed. There are a total of 18 segments in the lung on the left and right sides combined. Fifteen remaining segments divided by 18 equals 0.83; 0.83 times 1.1 equals 0.91 liters for a predicted postop FEV-1. Therefore, it would be considered acceptable to do the lobectomy if other patient factors including arterial blood gas are acceptable.

It is important to note that anytime a patient's postoperative FEV-1 will be marginal, a quantitative ventilation-perfusion scan is very useful. This will help determine which portions of the lung are nonfunctional and therefore do not contribute to overall lung function. For example, if a patient has a right upper lobe lesion and a marginal FEV-1, a ventilation-perfusion scan may be helpful because if there is little ventilation and little perfusion to the right upper lobe then this can be considered a nonfunctioning lobe and therefore resection will not hinder and, in fact, may help the patient's overall FEV-1. If the perfusion data and the ventilation data conflict as to the lung function, the perfusion data are always used.

In the example given above, if a ventilation-perfusion scan is done in this patient with a preoperative FEV-1 of 1.1 which reveals that 35% of the perfusion is going to the right upper lobe, then it can be presumed that 100-35 = 65% of remaining lung function will be present postoperatively; 65% x 1.1 = 0.72 liters for the predicted postoperative FEV-1; this precludes right upper lobectomy.

Another important point is that for any lung lesion, one must ask the following question before entering the operating room: Can the patient tolerate a pneumonectomy if this is needed? For example, if the patient has a right lower lobe lesion and is found to have a level 11 interlobar lymph node, then many thoracic surgeons will perform a pneumonectomy because the interlobar node is impinging on the right upper lobe as well. The patient may be able to tolerate a lobectomy but not a pneumonectomy based on the preoperative pulmonary functions, and therefore it is always important to keep in mind before doing the operation whether or not a pneumonectomy can be performed in any given patient.

OCCULT CANCER

12

A diagnosis of lung cancer sometimes occurs before it becomes radiologically apparent. A positive sputum cytology needs to be investigated for precise location and this includes careful head and neck examination, (since the positive cytology may actually be coming from the pharynx), bronchoscopy, or repeated diagnostic bronchial brushings from the same segmental bronchus. Occult cancers are almost always squamous cell carcinomas.

SEGMENTECTOMY VERSUS LOBECTOMY

The standard lung cancer operation for a tumor confined to a lobe of the lung is a lobectomy rather than a segmental resection. However, there is evidence that in some situations a segmentectomy may result in similar long-term survival as a lobectomy. This is important in a patient with limited pulmonary function in whom a lobectomy will result in a poor predicted postoperative FEV-1. In these segments a segmentectomy may permit resection of the lesion in an anatomical way but limit the amount of pulmonary compromise. In patients with squamous cell carcinoma, it appears that there is about equal initial benefit from a segmentectomy compared to a lobectomy; however, it appears that there is an overall 20% survival benefit in patients who undergo resection of the tumor as a lobectomy rather than a segmental resection. In particular, adenocarcinomas do poorly with limited resections, but more information is needed before any final conclusions are drawn about the overall applicability of segmentectomy.

RADIATION THERAPY

The rule of radiation therapy in stage I and II lung cancer is being debated. Even though radiation therapy decreases the recurrence rate in stage I and II carcinoma, this is not translated into improved survival. Therefore at this time for stage I and II nonsmall cell lung carcinoma, only resection is recommended and no additional radiation therapy. One exception is the superior sulcus tumors. Radiation is extremely useful preoperatively for superior sulcus tumors and has been shown to improve disease free survival. This is not the case for T3 lesions invading other portions of the chest wall, and these may simply require en bloc resection without preoperative radiation although this has not been fully defined. Radiation is useful for more advanced malignancies.

N2 DISEASE

Metastases to mediastinal lymph nodes are noted in nearly half of patients with nonsmall cell lung carcinoma. Although pessimism prevails about the outcome of treatment of N2 disease, a select group of patients with mediastinal node disease may benefit from resection. In a right-sided thoracotomy and pulmonary resection, fairly complete dissection of ipsilateral, mediastinal and subcarinal lymph nodes is possible. This is much more difficult in tumors involving the left lung. On the right side, the superior mediastinum and subcarinal region and inferior mediastinum are accessible. On the left side, the subaortic region is accessible, but the superior mediastinum is very difficult to get access to. The subcarinal region and inferior mediastinum can be approached from the left side, however. The prognostics and significance of positive mediastinal nodes differs if the mediastinal involvement with carcinoma is determined at the time of the mediastinoscopy or if it is determined at the time of thoracotomy. If the N2 disease is discovered at mediastinoscopy and a pulmonary resection done, only about 10% survive 5 years. On the other hand, if the N2 disease is previously unsuspected but is discovered at thoracotomy, approximately one-third survive 5 years. The implication is that patients found to have micrometastases to mediastinal nodes at the time of lung resection may have a good chance for survival if adequate mediastinal dissection and extirpation of tumor is performed.

T3 CHEST WALL INVOLVEMENT

Chest wall invasion by lung carcinoma occurs in approximately 5% with a 5 year post-resection survival rate of approximately 1 in 3 when regional lymph nodes are negative. There is general agreement that full thickness chest wall resection is indicated in patients with nonsmall cell lung cancer invading the chest wall. There is some disagreement if the tumor extension appears to involve only the parietal pleura. Some feel that an extrapleural dissection should be attempted and if a plane is readily achieved, then only an extrapleural resection should be performed without chest wall resection. However most feel that if the parietal pleura is involved then formal chest wall resection is indicated. The role of adjuvant radiation therapy has not yet been defined.

SUPERIOR SULCUS TUMORS

Superior sulcus tumors are a special type of lung carcinoma invading the chest wall. This is usually a low grade squamous cell carcinoma arising at the apex of the upper lobes which invades the endothoracic fascia, and may involve the lower roots of the brachial plexus, the sympathetic chain, and adjacent ribs and often vertebral bodies (Fig. 12.3a-d). The history and physical examination are important since these patients frequently develop characteristic pain in the shoulder, radiating down the arm in the ulnar distribution and may have the Horner's syndrome.

Superior sulcus tumors by definition are T3 disease (at least), and are stage IIIb if there is no nodal disease (N0), or stage IIIa if there is intrapulmonary node involvement (N1). The management of superior sulcus tumors is important since this is perhaps one of the few stage III types of lung cancer that may actually be cured by surgery.

Radiation alone has been reported to relieve pain and prolong survival. The diagnosis is established by sputum cytology and bronchoscopy, and if these cannot

12

Fig. 12.3a. CXR showing a left superior sulcus tumor (arrow).

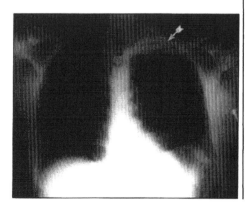

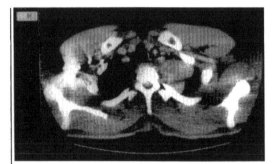

Fig. 12.3b. The CT confirms the superior tumor mass. The patient underwent left upper lobectomy with en bloc resection of the chest wall including ribs 1, 2 and 3.

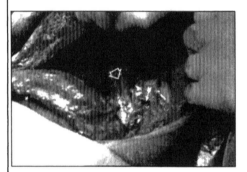

Fig. 12.3c. A portion of the T1 nerve root of the brachial plexus was invaded and required resection (arrow). The subclavian artery and vein are seen in the operative field (large arrow). A utility fifth interspace thoracotomy (*) facilitated the lobectomy.

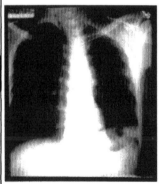

Fig. 12.3d. A postoperative CXR shows the resected portion of chest wall and absence of tumor.

12

establish the diagnosis then fine needle aspiration of the upper lobe lesion is performed. An assessment is made for resectability, both in terms of the patient as an operative candidate (pulmonary function tests and other co-morbid conditions are evaluated), as well as resectability in terms of the local tumor and possibility of metastases. The CT scan is invaluable for this and requires scanning of the local lesion and evaluation of surrounding invasion, including bony erosion of the vertebral bodies. Evaluation of mediastinal nodes is required since if these appear large then mediastinoscopy or mediastinotomy is essential to see whether there is mediastinal node involvement which would preclude resection. Also the CT scan should include evaluation of the upper abdomen to look at the liver and adrenal glands.

Preoperative radiation (3000 rads over 3 weeks or 4000 rads over 4 weeks in another protocol) followed by a 3 week period of rest and recuperation, followed by resection is standard therapy. The patient is placed in the lateral decubitus position with the arm raised. A parascapular incision is made starting above the spine of the scapula and extending around the tip of the scapula to the anterior axillary line. The 4th (or the 3rd) intercostal space is entered and an exploration is performed. The resection is done in two phases: anterior and posterior. The anterior phase involves division of the 2nd and 3rd ribs anterior to the tumor. The first rib is separated at the costochondral junction. The anterior end of the first rib is grasped and the scalenus anterior and medius are divided. The lower roots of the plexus, subclavian artery and subclavian vein are identified and carefully avoided. If tumor involves the lower roots of the brachial plexus, these may be divided. The posterior phase begins at the lower most divided rib and extends superiorly. Transverse processes and the sympathetic chain are taken posteriorly. The scalenus posterior is taken on the first rib. C8 is taken if it is involved. If C8 is not involved, then only T1 is taken. A portion of the vertebral body may be taken if needed (up to one-quarter of the vertebral body). There is no need for a prosthetic patch since support is provided by the scapula. An upper lobectomy is then performed en bloc with a resected chest wall.

BRAIN METASTASES

The incidence of brain metastases is higher in patients who have adenocarcinoma of the lung than in other histologic types of lung cancer. Untreated brain metastases have a mean survival of only a month and this is doubled if the patient is treated with steroids alone and improved to 4 months with cranial radiation. There are some instances that a solitary lesion in the brain may be resectable if there are no other contraindications. Therefore, in patients with nonsmall cell lung cancer who have good prognosis based on the management of their primary cancer, a solitary brain metastasis presenting simultaneously or after lung resection should be surgically excised. The role of postoperative radiation therapy to the brain is unclear at this point.

Other than solitary brain metastases, the presence of extrathoracic metastatic lesions for nonsmall cell lung carcinoma precludes resection of the primary tumor.

TECHNICAL CONSIDERATIONS

RIGHT PNEUMONECTOMY

A serratus anterior muscle sparing incision is performed in the 5th intercostal space using a posterolateral thoracotomy. The serratus sparing maneuver is optional since the serratus may be taken to improve exposure as necessary. Alternatively, some surgeons prefer to spare the latissimus dorsi and take the serratus anterior.

An exploratory thoracotomy is performed and an assessment is made for resectability. If there are miliary metastases throughout the parietal pleura, it can be presumed this is unresectable disease and biopsies are taken to establish the diagnosis. The lesion itself is examined carefully to ensure resectability. The mediastinal nodes are carefully examined. There should have been a preoperative evaluation by a chest CT scan to evaluate mediastinal nodes. If they were smaller than 1 to 1.5 cm in diameter, then thoracotomy is indicated for a primary lung lesion. Nonetheless, on exploratory thoracotomy, evaluation of the mediastinal nodes is mandatory.

If at the time of thoracotomy a positive mediastinal node is identified, excision of the primary lesion is still indicated since the thoracotomy has already been done and it has been shown that resection of the tumor and mediastinal node removal imparts a higher survival than if the diagnosis of mediastinal involvement is made preoperatively by mediastinoscopy. This implies a less advanced stage of the carcinoma when the diagnosis is made intraoperatively rather than preoperatively.

The general order of anatomic structures on the right is pulmonary vein most anteriorly, then pulmonary artery, then bronchus most posteriorly. The right mainstem bronchus is eparterial, i.e., it is superior to the level of the right pulmonary artery and is posterior to it.

The key relationship in performing a right pneumonectomy (as well as a right upper lobectomy) is developing a plane between the right superior pulmonary vein and the main pulmonary artery (Fig. 12.4a-e). The sequence of the operation is as follows:

1. The right superior pulmonary vein is ligated with silk ligature. It is usually necessary to ligate the tributaries of the right superior pulmonary vein separately. The pulmonary artery is now exposed. The right upper lobe pulmonary artery is taken and then the right main pulmonary artery is taken. It is usually not possible to take the right main pulmonary artery itself without taking the right upper lobe pulmonary artery separately. This is because the right pulmonary artery is shorter than the left and care must be taken that the tie on the right main pulmonary artery does not slip off.

2. Next the inferior pulmonary ligament and right lower lobe pulmonary vein are taken. The pulmonary vein is at the superior most aspect of the inferior pulmonary ligament.

3. Next the right mainstem bronchus which is posterior to the right pulmonary artery is taken with a TA bronchial stapling instrument. The lung is removed. A pleural flap is then placed over the bronchus or alternatively an intercostal muscle bundle. The main hazards with respect to a right pneumonectomy include uncontrolled hemorrhage from a

Fig. 12.4a. Right pneumonectomy. a) Exposure of right pulmonary hilum via postero-lateral thoracotomy. Reprinted with permission. Waldhausen JA, Pierce WS, Campbell DB. Pneumonectomy. In: Sabiston DC Jr, Spencer FC, eds. Surgery of the Chest. 6th ed. Mosby-Year Book, Inc., 1996:97,99.

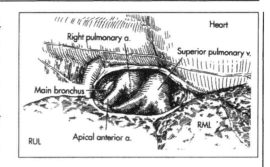

Fig. 12.4b. Right pulmonary artery taken. Reprinted with permission. Waldhausen JA, Pierce WS, Campbell DB. Pneumonectomy. In: Sabiston DC Jr, Spencer FC, eds. Surgery of the Chest. 6th ed. Mosby-Year Book, Inc., 1996: 97,99.

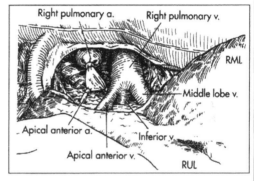

Fig. 12.4c. Superior pulmonary vein taken. Inferior pulmonary vein will be taken next. Reprinted with permission. Waldhausen JA, Pierce WS, Campbell DB. Pneumonectomy. In: Sabiston DC Jr, Spencer FC, eds. Surgery of the Chest. 6th ed. Mosby-Year Book, Inc., 1996:97,99.

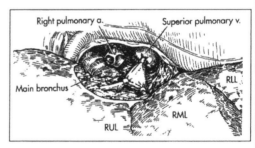

12

Fig. 12.4d. Right mainstem bronchus staple transected.

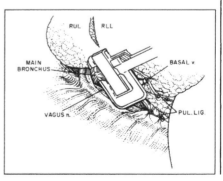

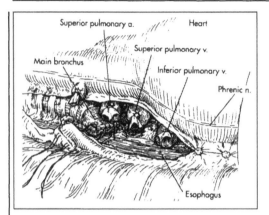

Fig. 12.4e. Hilum after lung is removed. Reprinted with permission. Waldhausen JA, Pierce WS, Campbell DB. Pneumonectomy. In: Sabiston DC Jr, Spencer FC, eds. Surgery of the Chest. 6th ed. Mosby-Year Book, Inc., 1996: 97,99.

poorly visualized right pulmonary artery, damage to the esophagus, and an excessively short right mainstem bronchus. In this case sleeve resection of the trachea may be needed.

LEFT PNEUMONECTOMY (Fig. 12.5)

Through the fifth intercostal space, a standard posterolateral thoracotomy is performed. The important features anatomically are: (1) the relationship with the left pulmonary artery to the aortic arch with reference to the ligamentum arteriosum and the left recurrent nerve; (2) the accessibility of the left pulmonary artery and both pulmonary veins intrapericardially; (3) inaccessability of the left mainstem bronchus since it is behind the arch of the aorta. The steps are as follows:

1. First the left superior pulmonary vein, which is located anteriorly, is ligated with silk.
2. The left pulmonary artery is taken with a vascular stapling instrument or with silk ties. Unlike the right pulmonary artery, it is rarely necessary

Fig. 12.5a. Left pneumonectomy. Exposure of left pulmonary hilum via posterolateral thoracotomy. Reprinted with permission. Vanecko RM, Neptune WB. Lung. In: Nora PF, ed. Operative Surgery. 3rd ed. Philadelphia: W.B. Saunders Co., 1990: 319.

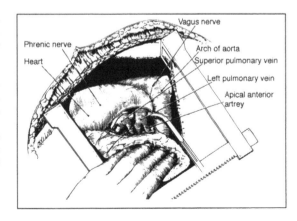

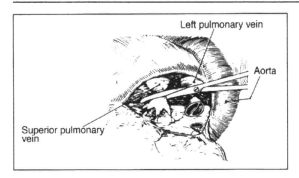

Fig. 12.5b. Left superior pulmonary vein and left pulmonary artery taken. The left inferior pulmonary vein will be taken next. Reprinted with permission. Vanecko RM, Neptune WB. Lung. In: Nora PF, ed. Operative Surgery. 3rd ed. Philadelphia: W.B. Saunders Co., 1990:319.

to take the left upper lobe pulmonary artery separately, since this first branch of the left pulmonary artery arises at quite a distance from the left pulmonary artery. On the right side, it is usually necessary to separately ligate the upper lobe arterial branch.

3. Next the inferior pulmonary ligament is taken down and the left inferior pulmonary vein is taken, usually by a vascular stapling instrument.
4. Next the left mainstem bronchus (which is located inferior and anterior to the left pulmonary artery (i.e., hyparterial)) is taken with a bronchial stapling instrument. A pleural flap is placed over the left mainstem bronchus to eliminate chances of a leak and to buttress over anastomosis.

Right Upper Lobectomy (Fig. 12.6a-f)

This is performed through either 4th or 5th intercostal space. Most surgeons prefer entering the 5th intercostal space through a posterolateral thoracotomy.

12

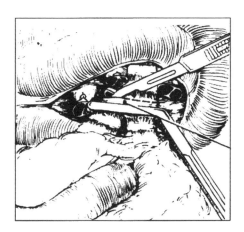

Fig. 12.5c. The left mainstem bronchus is transected. A staple gun is usually used prior to the transection. Reprinted with permission. Vanecko RM, Neptune WB. Lung. In: Nora PF, ed. Operative Surgery, 3rd ed. Philadelphia: W.B. Saunders Co., 1990:319.

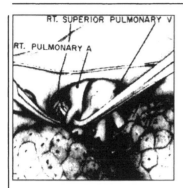

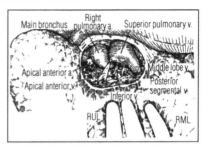

Fig. 12.6a. Right upper lobectomy. Pulmonary artery and vein identified. Reprinted with permission. Van Trigt P. Lung Infections and Diffuse Interstitial Lung Disease. In: Sabiston DC Jr, Spencer FC, eds. Surgery of the Chest. 5th ed. WB Saunders Co., 1989:652.

Fig. 12.6b. Right upper lobe pulmonary artery is taken and upper lobe pulmonary veins ligated. Reprinted with permission. Waldhausen JA, Pierce WS, Campbell DB. Pneumonectomy. In: Sabiston DC Jr, Spencer FC, eds. Surgery of the Chest. 6th ed. Mosby-Year Book, Inc., 1996:137.

The anterior surface of the hilum is dissected by incising the mediastinal pleura to expose the pulmonary veins. Again the most anterior structure is taken first, i.e., the right superior pulmonary vein is identified. The right upper lobe pulmonary vein is ligated. Care is taken to preserve the right middle lobe vein since injury to this may cause venous stasis and infarction of the right middle lobe. The right upper lobe pulmonary vein is then divided exposing the large anterior branch of the right pulmonary artery. The fissure is developed and the large truncus anterior branch of the right pulmonary artery is taken anterosuperiorly near the hilum. It should be noted that this is taken outside the fissure. The only branch to the right upper lobe which is taken in the fissure is the recurrent ascending posterior artery. This is a very important point since taking only the truncus anterior without entering the fissure will miss the recurrent ascending posterior branch and could be torn when removing the right upper lobe. The right upper lobe bronchus is the final structure and is taken with a TA-30 4.8 mm staple gun.

Next the right middle lobe may be plicated to the right lower lobe with 2-0

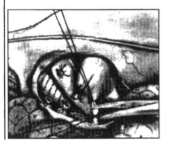

Fig. 12.6c. Upper lobe pulmonary veins are cut. Care is taken to avoid injury particularly to the right middle lobe vein. Reprinted with permission. Van Trigt P. Lung Infections and Diffuse Interstitial Lung Disease. In: Sabiston DC Jr, Spencer FC, eds. Surgery of the Chest. 5th ed. WB Saunders Co., 1989:653.

12

Fig. 12.6d. Interlobar fissure developed. Reprinted with permission. Van Trigt P. Lung Infections and Diffuse Interstitial Lung Disease. In: Sabiston DC Jr, Spencer FC, eds. Surgery of the Chest. 5th ed. WB Saunders Co., 1989:653.

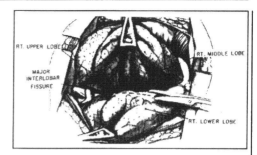

Fig. 12.6e. The posterior ascending artery is identified within the fissure and taken. Reprinted with permission. Van Trigt P. Lung Infections and Diffuse Interstitial Lung Disease. In: Sabiston DC Jr, Spencer FC, eds. Surgery of the Chest. 5th ed. WB Saunders Co., 1989:653.

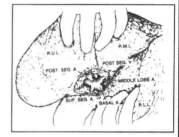

Fig. 12.6f. Right upper lobe bronchus identified from behind and staple-resected.

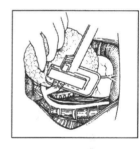

12

chromic stitches to prevent torsion. It should be noted that mobilization of the inferior pulmonary ligament is critical since it permits the remaining lung to be better able to occupy the remaining pleural space.

MIDDLE LOBECTOMY (Fig. 12.7A-E)

This is approached again via the standard lateral decubitus position. A posterolateral thoracotomy is employed, again entering the 5th interspace using a serratus sparing incision if possible. The inferior pulmonary ligament is mobilized. The mediastinal pleura is incised to identify and ligate the right middle lobe vein. This is a branch of the right superior pulmonary vein.

Next the right middle lobe pulmonary artery is identified and taken. One must be careful to avoid the posterior recurrent ascending branch to the upper lobe as well as the superior segmental branch to the lower lobe. All these vessels come off very close to one another. Finally the right middle lobe bronchus is taken. This is found beneath the right pulmonary artery. Again care must be taken to avoid

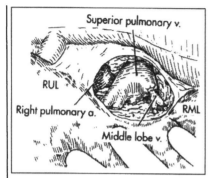

Fig. 12.7a. Right middle lobectomy. Right middle lobe pulmonary vein taken. Reprinted with permission. Waldhausen JA, Pierce WS, Campbell DB. Pneumonectomy. In: Sabiston DC Jr, Spencer FC, eds. Surgery of the Chest. 6th ed. Mosby-Year Book, Inc., 1996:139.

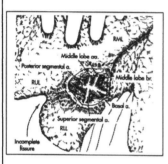

Fig. 12.7b. Middle lobe pulmonary artery taken within the fissure, avoiding the ascending posterior and superior segmental arteries. Reprinted with permission. Waldhausen JA, Pierce WS, Campbell DB. Pneumonectomy. In: Sabiston DC Jr, Spencer FC, eds. Surgery of the Chest. 6th ed. Mosby-Year Book, Inc., 1996:139.

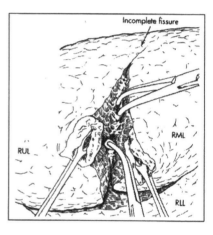

Fig. 12.7c. An incomplete fissure is further developed. Reprinted with permission. Waldhausen JA, Pierce WS, Campbell DB. Pneumonectomy. In: Sabiston DC Jr, Spencer FC, eds. Surgery of the Chest. 6th ed. Mosby-Year Book, Inc., 1996:139.

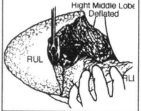

Fig. 12.7d, left. The middle lobe bronchus is staple transected. (12.7e, right) The right lung is inflated except for the middle lobe, which is then staple-resected.

Fig. 12.8a. Right lower lobectomy. Inferior pulmonary ligament mobilized. Reprinted with permission. Van Trigt P. Lung Infections and Diffuse Interstitial Lung Disease. In: Sabiston DC Jr, Spencer FC, eds. Surgery of the Chest. 5th ed. WB Saunders Co., 1989:658.

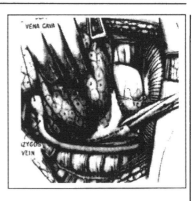

Fig. 12.8b. Right inferior pulmonary vein taken. Reprinted with permission. Van Trigt P. Lung Infections and Diffuse Interstitial Lung Disease. In: Sabiston DC Jr, Spencer FC, eds. Surgery of the Chest. 5th ed. WB Saunders Co., 1989:658.

injury to the superior segmental bronchus which comes out very close to the right middle lobe orifice.

RIGHT LOWER LOBECTOMY (**Fig. 12.8A-D**)

The right lower lobectomy is performed through the 6th intercostal space. The inferior pulmonary ligament is divided up to the level of the inferior pulmonary vein; this is then taken, either with ligatures or preferably now with a TA vascular stapling instrument. The right lower lobe pulmonary artery is then exposed to the fissure. This superior segmental artery is taken. This is at the level of the right middle lobe pulmonary artery and care must be taken not to injure this artery. Then the basal portion of the right lower lobe pulmonary artery is identified and taken. The right lower lobe bronchus is exposed and the superior segmental bronchus is carefully identified to avoid narrowing of the right middle lobe bronchus when taking the right lower lobe bronchus with a stapler.

As one can see from the above discussion, the general plan of the procedure is generally the same for all lobes. Sequentially, the vein, then artery, then bronchus

Fig. 12.8c. The pulmonary artery is identified in the fissure and the superior segmental artery and basal artery are taken separately. Reprinted with permission. Waldhausen JA, Pierce WS, Campbell DB. Pneumonectomy. In: Sabiston DC Jr, Spencer FC, eds. Surgery of the Chest. 6th ed. Mosby-Year Book, Inc., 1996:145.

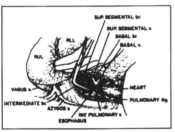

Fig. 12.8d. The right lower lobe bronchus is stapled from behind. Reprinted with permission. Waldhausen JA, Pierce WS, Campbell DB. Pneumonectomy. In: Sabiston DC Jr, Spencer FC, eds. Surgery of the Chest. 6th ed. Mosby-Year Book, Inc., 1996:145.

are divided in that order. Alternatively, one may take the artery first to allow for pulmonary devascularization prior to removal.

LEFT UPPER LOBECTOMY (Fig. 12.9A-E)

This is done via a posterolateral thoracotomy through the 4th or 5th intercostal space. The left upper lobe is swept down from the apex of the left pleural space. The left upper lobe pulmonary artery is taken as it curves over the left upper lobe bronchus. Notice that on the left side the bronchus is hyparterial, i.e., it is inferior and anterior to the left pulmonary artery. It is easiest to identify the anterior and

Fig. 12.9a. Left upper lobectomy. The left upper lobe is retracted inferiorly; pleura is incised between the left pulmonary artery and the left upper lobe. Reprinted with permission. Van Trigt P. Lung Infections and Diffuse Interstitial Lung Disease. In: Sabiston DC Jr, Spencer FC, eds. Surgery of the Chest. 5th ed. WB Saunders Co., 1989:661.

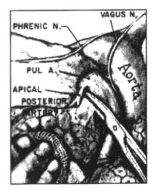

Fig. 12.9b. Apical-posterior and anterior branches are taken from this superior approach. Reprinted with permission. Van Trigt P. Lung Infections and Diffuse Interstitial Lung Disease. In: Sabiston DC Jr, Spencer FC, eds. Surgery of the Chest. 5th ed. WB Saunders Co., 1989:661.

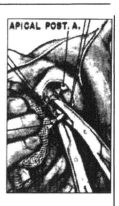

Fig. 12.9c. Interlobar fissure is developed and lingular arteries taken within the fissure. Reprinted with permission. Van Trigt P. Lung Infections and Diffuse Interstitial Lung Disease. In: Sabiston DC Jr, Spencer FC, eds. Surgery of the Chest. 5th ed. WB Saunders Co., 1989:661

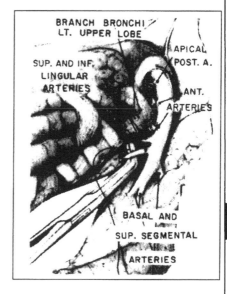

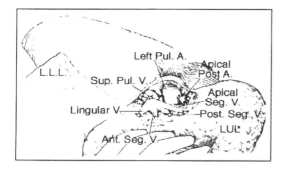

Fig. 12.9d. Left superior pulmonary vein taken anteriorly. Reprinted with permission. Waldhausen JA, Pierce WS, Campbell DB. Pneumonectomy. In: Sabiston DC Jr, Spencer FC, eds. Surgery of the Chest. 6th ed. Mosby-Year Book, Inc., 1996:129.

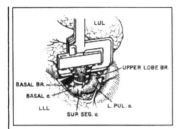

Fig. 12.9e. Left upper lobe bronchus staple-transected from behind.

apical-posterior segmental arteries of the left upper lobe pulmonary artery in the base of the fissure. The lingula branch is then identified. All these branches are then ligated, and the left upper lobe pulmonary artery is ligated proximally.

As mentioned before, it may be possible on the left side to simply ligate the left upper lobe pulmonary artery proximally and distally, since this artery is longer than on the right side. The left superior pulmonary vein is then taken, preferably with a TA vascular stapling device. The left upper lobe bronchus is then taken with the TA-30 4.8 mm stapling device.

Going back to the separation of the arteries, one needs to separately take the anterior, apical-posterior and lingular arteries. The posterior and apical arteries are taken anterosuperiorly outside the fissure; the lingula artery and anterior arteries are taken from the fissure aspect.

LEFT LOWER LOBECTOMY (**Fig. 12.10**A-C)

This is similar in nearly all respects to right lower lobectomy. It is done through a 6th intercostal space via a posterolateral thoracotomy. The inferior pulmonary vein is taken at the superiormost aspect of the inferior pulmonary ligament. The left lower lobe pulmonary artery is then taken in the fissure near the takeoff of the lingular artery. The left lower lobe bronchus is then taken by a stapling device being sure to avoid the left upper lobe bronchus which is close to the take-off of the superior segmental bronchus of the lower lobe. Clear definition is important to avoid injury of the left upper lobe bronchus.

SEGMENTECTOMY

Although all segments of the right upper lobe may be taken by segmentectomy, this is rarely done because the segmental bronchi are usually short and very difficult to take separately. Therefore, generally segmental right upper lobe resections are not performed. Similarly right middle lobe resections are generally not performed. For the right lower lobe, right superior segmentectomies may be readily performed as may right basilar segmentectomies. On the left side for the left upper lobe, a left apical posterior segmentectomy may be performed; likewise, a left anterior segmentectomy may be performed. A lingular segmentectomy may also be performed. On the left side, an apical or basal segmentectomy may again be performed for the lower lobe.

Overall, segmentectomy is most often done for the superior segments of the lower lobes. The next most common segmentectomy is the lingulectomy. These will now be discussed.

Fig. 12.10a. Left lower lobectomy. a) Left inferior pulmonary ligament mobilized and left inferior pulmonary vein taken. Reprinted with permission. Waldhausen JA, Pierce WS, Campbell DB. Pneumonectomy. In: Sabiston DC Jr, Spencer FC, eds. Surgery of the Chest. 6th ed. Mosby-Year Book, Inc., 1996:131.

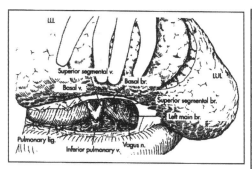

Fig. 12.10b. The pulmonary artery is identified in the fissure and the superior segmental and basal branches are taken. Reprinted with permission. Waldhausen JA, Pierce WS, Campbell DB. Pneumonectomy. In: Sabiston DC Jr, Spencer FC, eds. Surgery of the Chest. 6th ed. Mosby-Year Book, Inc., 1996:131.

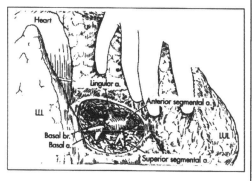

Fig. 12.10c. The left lower lobe bronchus is staple-transected and the lower lobe removed.

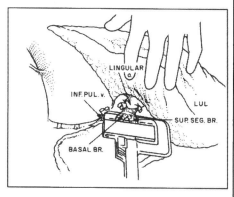

12

For the superior segmentectomy of either side, first the arterial and venous supplies of the superior segment are taken. This includes careful dissection to avoid injury to the right middle lobe artery on the right and the lingular artery on the left within the fissure. Likewise, the right middle lobe vein and the left lingular vein need to be carefully identified to avoid injury to the left and right superior pulmonary vein. After the vessels are taken, the superior segmental bronchus is clamped, being careful to avoid the right middle lobe orifice or the left upper lobe orifice. With the superior segmental bronchus clamped, the lung is inflated. The noninflated portion of the lung represents the superior segment which is then staple-resected with a GIA or TA instrument.

A similar overall procedure is carried out when performing a basal segmentectomy on either side, except that the superior segmental vessels and bronchus are preserved and it is the basal segmental vessels and bronchus that are taken. Similarly, the basal segmental bronchus is clamped and when inflating, it is the basal area that will remain noninflated; it is this that is staple-resected.

For a lingular segmental resection, the lingular artery is controlled in the fissure. The lingular vein is then taken. Finally, the lingular bronchus is clamped, the lung inflated and that portion of the lung not inflating (corresponding to the lingula) is staple-resected.

Generally, segmentectomy is an inferior cancer operation to a lobectomy, although a segmentectomy is more anatomic and therefore a superior operation to a wedge resection for neoplasia. Nonetheless, wedge resections and segmentectomies have definite roles in patients who have limited pulmonary reserve and in whom a lobectomy would not be safe because of poor predicted postoperative pulmonary functions.

SLEEVE RESECTION

A sleeve resection is a localized resection of a bronchial orifice with a bronchoplasty or reimplantation of the distal bronchial segment to the proximal bronchial segment. This is shown in Figures 12.11 and 12.12a-c. Those lesions which are amenable to sleeve resection are a tumor in the orifice of the right upper lobe bronchus; a tumor at the orifice of the middle lobe bronchus; a tumor at the orifice of the left upper lobe bronchus or at the orifice of the left lower lobe bronchus. The most common indication for a sleeve resection is a carcinoid tumor at the orifice of the right upper lobe bronchus. In this case, a right upper lobectomy can be performed with resection of the orifice of the right upper lobe and the distal portion of the right mainstem bronchus, as well as the proximal portion of the bronchus intermedius. The bronchus intermedius is then anastomosed to the right mainstem bronchus as a bronchoplasty. The advantage is that the right middle and right lower lobes are preserved. More malignant tumors at the origin of the right upper lobe bronchial orifice may likewise undergo bronchoplasty. Without a bronchoplasty sleeve resection procedure, a pneumonectomy would be required for such a lesion. The technical details of a right upper lobe sleeve resection will be described.

A right 4th to 5th intercostal space posterolateral thoracotomy is performed and assessment is made for resectability, i.e., critical search for mediastinal nodes, interlobar nodes or diffuse metastases, all of which would preclude a sleeve resection. The inferior pulmonary ligament is taken. The hilum is mobilized and the right upper lobe branch of the right superior pulmonary vein is taken. Then either anteriorly or within the fissure, the right superior pulmonary artery (truncus anterior) is taken and the posterior ascending artery is separately taken within the fissure. Posteriorly, the take-off of the right upper lobe bronchus from the right main bronchus is identified. The right mainstem is transected medially, the bron-

Fig. 12.11. Tumor locations for which sleeve resections are technically feasible.

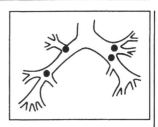

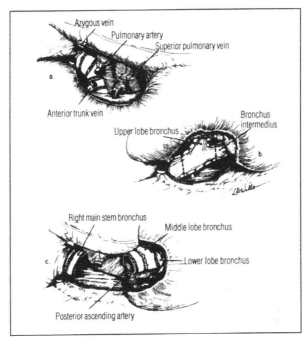

Fig. 12.12. Right upper lobectomy and sleeve resection of adjacent mainstem bronchus. a) Ligation of azygous vein and right upper pulmonary artery and vein branches. b) Proposed sleeve resection. c) Right upper lobe has been resected. A bronchoplasty between the right mainstem bronchus and the bronchus intermedius will be done. Reprinted with permission. Shields TW. Bronchial Sleeve Lobectomy. In: Shields TW, ed. General Thoracic Surgery. 3rd ed. Lippincott Williams & Wilkins, p. 379.

chus intermedius is transected proximally, and the right upper lobe is removed en bloc. The bronchus intermedius is then anastomosed to the right mainstem bronchus with interrupted 3-0 Vicryl. A pleural or pericardial flap is placed over the bronchoplasty to increase vascularity and provide support in the event of a leak.

Postoperative pulmonary care must be meticulous and bronchoscopy is warranted immediately with any evidence of right middle or lower lobe collapse or infiltrate.

SUGGESTED READING

1. Bains MS. Surgical treatment of lung cancer. Chest 1991; 100:826-837.

2. Aisner J, Alberto P, Bitran J et al. Role of chemotherapy in small cell lung cancer: A consensus report of the International Association for the study of lung cancer workshop. Cancer Treat Ref 1983; 67:37-43.

3. Meyer JA. Five year survival in treated stage I and II small cell carcinoma of the lung. Ann Thorac Surg 1986; 42:668-669.

4. Pearson FG, Nelems JM, Henderson RD et al. The role of mediastinoscopy in the selection of treatment for bronchial carcinoma with involvement of superior mediastinal lymph nodes. J Thorac Cardiovasc Surg 1972; 64:382-390.

5. Pancoast HK. Superior pulmonary sulcus tumor: tumor characterized by pain, Horner's syndrome, destruction of bone and atrophy of hand muscles. JAMA 1932; 99:1391-1396.

6. Paulson DL. Superior sulcus carcinomas. In: Sabiston DC Jr, Spencer FC, eds. Gibbon's surgery of the chest. 4th ed. Philadelphia: WB Saunders, 1983:506.

7. Grillo HC, Zannini P, Mickelassi F. Complications of tracheal reconstruction: incidence, treatment, and prevention. J Thorac Cardiovasc Surg 1986; 91:322-328.

8. Wilkins EW Jr. The asymptomatic isolated pulmonary nodule. N Engl J Med 1955; 52:515-520.

9. Mountain CF. Revisions in the international system for staging lung cancer. Chest 1997; 111:1710.

10. Mountain CF, Dresler CM. Regional lymph node classification for lung cancer staging. Chest 1997; 111:1718.

11. McCaughan BC et al. Chest wall invasion in carcinoma of the lung: Therapeutic and prognostic implications. J Thorac Cardiovasc Surg 1985; 89:836.

12. Mansour KA, Wanna FS. Extended resection of bronchial carcinoma in the superior pulmonary sulcus. In: Shields TW, LoCicero J III, Ponn RB, eds. Philadelphia: Lippincott Williams & Wilkins, 2000:467-72.

12

Empyema

Fritz J. Baumgartner

The management of empyema can be a difficult and often confusing subject. There are several fundamental questions that must be answered (Table 13.1). Is it acute or chronic? Is there a space problem which requires obliteration? Is there a bronchopleural fistula? Is this a postoperative problem? The individual management depends on the answers to these questions.

The etiology of empyema is most commonly a result of lung abscess or pneumonia. The second most common etiology is post-surgical after lobectomy or pneumonectomy. Other causes include spontaneous pneumothorax, retained foreign body, or a subdiaphragmatic process causing supradiaphragmatic empyema.

NONPOST-RESECTION EMPYEMA

For nonpost-resection empyema, i.e., that empyema arising most commonly as a result of a lung abscess or pneumonia, thoracentesis is the procedure done first (Table 13.2). Either clear water fluid or pus is produced. For clear watery fluid, cultures and chemistries are obtained. If there is a positive culture, a chest tube is placed. For negative culture, and if it is a transudate by chemistry evaluation, then the patient only will require a repeat thoracentesis if the effusion reaccumulates. If, however, it is an exudate by chemistries (i.e., if the specific gravity is greater than 1.02, the white blood count greater than 500 per cubic millimeter, the protein greater than 2.5 grams/dl, the pH less than 7.00, the glucose less than 40 mg/dl, or the LDH greater than 1000, then a chest tube is placed. If pus is obtained from the initial thoracentesis, then a chest tube is placed and this chest tube is applied to suction drainage. After about 10 days, this is converted to open drainage via an empyema tube. A sinogram is then obtained. If no cavity is present, then the chest tube can be slowly withdrawn over a period of several days to weeks. If a small cavity is found and if the chest tube is draining well, then again the tube may be slowly withdrawn. If the cavity is not well-drained by the chest tube, a new chest tube is placed or a rib is resected to facilitate drainage. If there is a large cavity which appears to be well-drained, then the tube is slowly withdrawn. If the lung does not re-expand fully, then decortication is indicated between 6 weeks to 8 weeks later. If there is a large cavity which is not well-drained, rib resection is indicated to facilitate drainage or a Clagett window or Eloesser flap is performed. A Clagett window involves resection of several ribs and marsupialization of the skin such that the skin is tacked down to the intercostal muscles permitting good drainage. The physician/nurse and ultimately the patient can irrigate and pack

13

Table 13.1. Empyema

- Is there adequate drainage?
- Is there a space problem?
- Is there a bronchopleural fistula?
- Is this postop?
- Is this acute or chronic?

Drainage procedures
- Chest tube
- Rib resection
- Clagett window/Eloesser flap

Airspace filling procedures
- Thoracoplasty
- Fill space with muscle and omentum

Bronchopleural fistula after surgery
- Early reoperation
- Transpericardial approach to close fistula
- Muscle flap over fistula

Table 13.2. Empyema-nonpost-resection

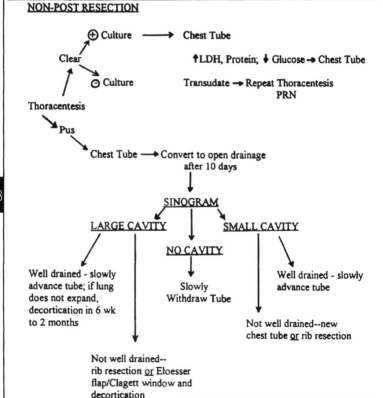

this wound. The Eloesser flap is a specific form of drainage in which a tube of skin is literally laid down into the cavity to facilitate drainage. It is probably considered less and less in this day and age, since rib resection and Clagett window have served so well. At the time of drainage of the empyema a decortication of the lung is also performed.

POST-RESECTION EMPYEMA

The management of post-resection empyema is shown in Table 13.3.

This is much more common after pneumonectomy than after lobectomy. If a post-resection empyema occurs, one must consider if there is a bronchopleural fistula present. If there is no bronchopleural fistula present, then chest tube drainage is employed and 2 weeks later it is converted to open drainage.

In the case of post-pneumonectomy empyema, the protocol is a little bit different. Chest tube drainage is employed for 2 weeks and then converted to open drainage under fluoroscopic guidance. If the mediastinum is frozen under fluoroscopic guidance, then it is safe to do a Clagett window for good drainage of the post-pneumonectomy empyema. If, however, the mediastinum seems to shift towards the side of the empyema after a pneumonectomy, then it is not yet safe to do a Clagett window, and the chest tube is placed back to closed drainage. There should not be suction applied at this point, lest the mediastinum be further shifted to the empyema side.

For post-resection empyema other than pneumonectomy and if there is no bronchopleural fistula present, chest tube drainage with conversion to open drainage may be all that is necessary. However, if there is persistent sepsis or evidence of problems, then a Clagett window may be necessary for better drainage, and a myoplasty/omentoplasty may be necessary to obliterate the dead space. Thoracoplasty may be warranted for post-resection empyema, particularly in the event that this is status post pneumonectomy. If there is a bronchopleural fistula present, then the protocol changes. If this is an early empyema and bronchopleural fistula, i.e., within 6 days postoperatively, chest tube drainage is indicated followed by immediate reoperation to repair the technical error. The plan is to go into the old incision, fix the bronchus and place a pleural flap or intercostal muscle flap over the repair. For late post-pneumonectomy empyema with bronchopleural fistula, again chest tube drainage is the initial procedure. A Clagett window is then performed early on to drain the empyema, and a simultaneous muscle flap procedure may be done. This should probably be done within a week after placement of the chest tube. If there is still persistent air leak despite the Clagett window and muscle flap procedure, then much later another attempt at repair of the bronchial stump can be done. If this was a pneumonectomy, then transsternal stapling of the bronchial stump may be indicated. The reason for doing a transsternal approach to the mainstem bronchus is that it permits access to virgin territory not affected by the empyema. The location of the carina after performing a median sternotomy is in the region just to the right of the ascending aorta, just to the left of the superior vena cava, and just inferior to the left innominate vein.

Table 13.3. Post-resection empyema

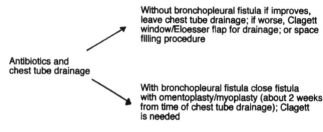

Antibiotics and
chest tube drainage

Without bronchopleural fistula if improves,
leave chest tube drainage; if worse, Clagett
window/Eloesser flap for drainage; or space
filling procedure

With bronchopleural fistula close fistula
with omentoplasty/myoplasty (about 2 weeks
from time of chest tube drainage); Clagett
is needed

(EMPYEMA WITH LOBECTOMIES CAN USUALLY BE MANAGED
SIMPLY WITH CHEST TUBE AND <u>MAYBE</u> RIB RESECTION)

POST-PNEUMONECTOMY EMPYEMA

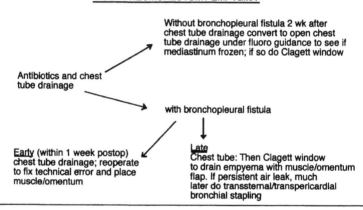

Antibiotics and chest
tube drainage

Without bronchopleural fistula 2 wk after
chest tube drainage convert to open chest
tube drainage under fluoro guidance to see if
mediastinum frozen; if so do Clagett window

with bronchopleural fistula

<u>Early</u> (within 1 week postop)
chest tube drainage; reoperate
to fix technical error and place
muscle/omentum

<u>Late</u>
Chest tube: Then Clagett window
to drain empyema with muscle/omentum
flap. If persistent air leak, much
later do transsternal/transpericardial
bronchial stapling

The most common cause of late bronchopleural fistula after pneumonectomy
is infection; less common is recurrent carcinoma, even as long as 2 years later. The
most common cause of early bronchopleural fistula is technical error, and this is
why repair immediately after an early bronchopleural fistula is identified is indicated.

For space filling procedures, a latissimus dorsi muscle flap is the most commonly used and most versatile muscle for reconstruction. Entry into the chest is
obtained by excising ribs 6 to 7 at the mid-lateral position. The pectoralis muscle
flap is achieved by resecting several centimeters of the 2nd and 3rd ribs at the
mid-lateral position. It should also be noted that empyema after lobectomy can
usually be managed simply with chest tube and perhaps rib resection for drainage. Lower lobes have larger residual space, and so there is an increased risk of
empyema with lower lobectomies. On occasion, Clagett procedures and myoplasty/

omentoplasty may be necessary.

The therapy for post-resection empyema can be classified as drainage procedures, air space filling procedures and management of bronchopleural fistula after surgery. The drainage procedures can be chest tube, rib resection or Clagett/ Eloesser flap. Air space filling procedures include thoracoplasty which is basically resection of ribs to permit collapse of the chest wall to obliterate the dead space. Another air space filling procedure is, as mentioned, filling the space with muscle, omentum or even pericardium. For bronchopleural fistula after surgery, early reoperation is indicated if found early on. For later bronchopleural fistula, muscle flaps may be used over the fistula, and if all else fails, a paratranssternal transpericardial approach may be employed to close the fistula.

SUGGESTED READING

1. Lemmer JH, Botham JH, Orringer MB. Modern management of adult thoracic empyema. J Thoracic Cardiovasc Surg 1985; 90:849-855.
2. Clagett OT, Geraci JE. A procedure for the management of post-pneumonectomy empyema. J Thorac Cardiovasc Surg 1963; 45:141.
3. Hankins JR et al. Bronchopleural fistula: Thirteen year experience with 77 cases. J Thorac Cardiovasc Surg 1978; 76:755.
4. Lai I, Unruh H. Management of empyema thoracis. Ann Thorac Surg 1990; 50:355.

13

Tuberculosis

Fritz J. Baumgartner

PULMONARY TUBERCULOSIS

Approximately 7% of the population is infected with the tubercle bacillus. In the upcoming years with more and more of the population becoming immuno-deficient because of HIV or neoplasia, it can be expected that tuberculosis will once again become a disease that will be in the forefront of thoracic surgery. About 10% of those infected with *Mycobacterium tuberculosis* will develop clinically sig-nificant disease. The clinical manifestations of tuberculosis are present more of-ten in the elderly and in new immigrant groups. *Mycobacterium tuberculosis* is so virulent that infection may be initiated by a single organism to an alveolus in susceptible people.

The bacilli are airborne and result in an initial infection of the lung which most often results in a primary complex (the Ghon's tubercle) with secondary foci of tuberculosis in the hilar lymph nodes. This results in a hypersensitivity reaction to the organism manifested by a positive PPD skin test defined as an area of induration and erythema greater than 1 cm diameter within 48 hours.

In the primary infection, the PPD skin test is negative for 8 weeks after the infection and subsequently becomes positive when the hypersensitivity reaction comes into effect. As mentioned, only 10% of these develop clinically significant disease and 90% are confined as a Ghon complex. The 10% to advance include the following four clinical manifestations:

1. Pleural effusion. If the patient has a positive PPD skin test and has a pleural effusion, then the chance of *Mycobacterium tuberculosis* as the source is 90%.
2. Miliary tuberculosis. Here there are multiple pulmonary nodules and the patients are usually anergic with a negative TB skin test. Half of these patients will die of this disease.
3. Tuberculosis pneumonia. Tuberculosis pneumonia results from aero-solization into the lung parenchyma; the TB skin test is typically positive.
4. Extrapulmonary latent disease. This may involve multiple organ sys-tems including kidney, brain, epididymis, liver or joint surfaces. The knee, bladder, appendix, cecum or terminal ileum are most commonly involved.

Tuberculosis in the chest should be considered in a similar fashion to appendi-citis in the abdomen, i.e., it should always be kept in the back of one's mind and

14

Fig. 14.1. Caseous lobular *M. tuberculosis* pneumonia with progressive abscess formation. Courtesy Dr. Alessandro DeVito, Harbor-UCLA.

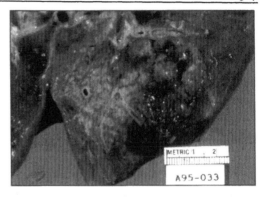

usually should be no less than second on the list of differential diagnoses for a chest process. Tuberculosis is the great imitator, just as appendicitis is a great imitator in the abdomen. It is a common disease that presents frequently with uncommon findings and should never be underestimated.

Indications for surgery for tuberculosis include pleural effusion in which drainage of the empyema is indicated as well as pleuroscopy and biopsy for diagnosis if necessary. Massive amounts of bleeding or pulmonary destruction found with TB pneumonia (Fig. 14.1) or miliary TB are also indications for surgery. Also, extrapulmonary manifestations in various organ systems may require surgical intervention.

It should be noted that another type of tuberculosis is becoming more and more prevalent in the immunocompromised population. This includes *Mycobacterium kansasii* and *intracellulare-avium*. In these patients, cavitation with multiple thin-walled cavities are more common than with *Mycobacterium tuberculosis*; however pleural effusions are rare. It is an important observation that patients with these atypical Mycobacteria are often clinically less sick than their terrible x-ray picture would suggest.

The x-ray manifestations of primary *Mycobacterium tuberculosis* may appear in the pulmonary parenchyma, in the hilar or mediastinal lymph nodes, or in the pleural space. Parenchymal involvement is most often in the mid-zone of the lung and looks like a pneumonia. Cavitation is not common in primary tuberculosis. Hilar or paratracheal lymph node enlargement may occur in adults, but is much more frequent in children and occurs in nearly all. Effusion on chest x-ray is more common in the adult than in the child. With post-primary pulmonary tuberculosis, i.e., reactivation tuberculosis, the disease is localized primarily to the apical and posterior segments of the upper lobes and the superior segments of the lower lobes, but other areas may be involved as well. The process consists of foci of caseous necrosis which may coalesce, liquefy and empty into a bronchus. Although the findings on x-ray are often characteristic, the diagnosis of tuberculosis can only be made by culture or microscopic examination of the acid-fast stain (Fig. 14.2). The mainstay of treatment of tuberculosis is chemotherapy. Commonly used agents include INH (isoniazid), rifampin and ethambutol. streptomycin, PZA

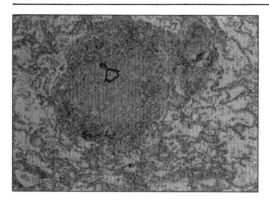

Fig. 14.2. Photomicrograph of a *M. tuberculosis* caseating granuloma with a Laughan's giant cell (arrow). The periphery of the granuloma is surrounded by lymphocytes and fibroblasts. Courtesy Dr. Alessandro DeVito, Harbor-UCLA.

(pyrazinamide) and ethionamide are used less commonly.

The indications for surgery in tuberculosis include destroyed lung, hemoptysis, rupture of an abscess into a pleura or into a bronchus, a bronchopleural fistula, inability to rule out carcinoma, and refractoriness to standard medical management with persistent positive sputum even after 6 weeks of adequate medical management. The atypical mycobacteria is generally more refractory to medical management than the *Mycobacterium tuberculosis*.

THORACOPLASTY

This was a primary operation for tuberculosis in the past, the philosophy being that decrease in the dead space resulted in collapse of the lung and prevention of growth of the tuberculosis. The indication now is for chronic empyema with excess dead space. It is a rarely performed procedure, nonetheless. Ordinarily, seven ribs are resected, not including the first rib. The incision is a long parascapular incision from the spine of the scapula and extending to the tip of the scapula. Ribs are resected subperiosteally from the transverse process anteriorly to the costal cartilage anteriorly. The immediate physiologic changes include increased effort of breathing due to paradoxic motion. The cough mechanism is reduced because of the unsupported chest. Scoliosis long-term can result in further decrease in pulmonary function.

MASSIVE HEMOPTYSIS

Tuberculosis is the most common cause of massive hemoptysis which is defined as the production of 600 cc of blood within 24 h. In approximately three quarters of cases, tuberculosis is involved, and most of the rest are carcinoma-related. Massive, life-threatening hemoptysis is associated with tuberculosis with or without superimposed fungal infection, Aspergillus being the most common. The usual etiology of massive hemoptysis is erosion into a bronchial artery, or less commonly pulmonary arterial bleeding in cavitary tuberculosis resulting from a ruptured Rasmussen's aneurysm.

Emergency management of massive hemoptysis is required. The first emergency step is protecting the airway and adequacy of ventilation of uninvolved lung. This involves patient positioning to minimize aspiration of blood. For example, if the pathology is known to be in the right upper lobe, the patient is placed in the right lateral decubitis position. Bronchoscopy usually is needed to identify the source of hemoptysis, and then either endobronchial tamponade with isolated Fogarty balloon occlusion or double-lumen endotracheal intubation is established to protect the uninvolved airway. At this point, management of the bleeding usually proceeds using bronchial arterial embolization. In most patients short-term control of hemorrhage is achieved by embolization, and half achieve long-term control by embolization alone. Nonetheless, in the presence of severe tuberculous lung destruction, with or without fungal invovlement, definitive surgical resection usually is required. Using preoperative temporizing angiographic embolization, definitive pulmonary resection can be done on a more elective basis to optimize the patient as a surgical candidate.

SUGGESTED READING

1. Churchill ED, Klopstock R. Lobectomy for pulmonary tuberculosis. Ann Surg 1943; 117:641-669.
2. Shields TW. Pulmonary tuberculosis and other mycobacterial infections of the lung. In: Shields TW, ed. General Thoracic Surgery. 3rd ed. Philadelphia: Lea and Febinger, 1989:785-798.
3. Treasure RL, Seaworth BJ. Current role of surgery in mycobacterium tuberculosis. Ann Thorac Surg 1995; 59:1405.
4. Pomerantz M et al. Surgical management of resistant mycobacterium tuberculosis and other mycobacterial pulmonary infections. Ann Thorac Surg 1991; 52:1108.
5. Mouroux J et al. Surgical management of pleuropulmonary tuberculosis. J Thorac cardiovasc Surg 1996; 111:662.
6. Uflacker R et al. Management of massive hemoptysis by bronchial artery embolization. Radiology 1983; 146:627.
7. Crocco JA, Rooney JJ, Fankuahen DS et al. Massive hemoptysis. Arch Intern Med 1968; 121:495-498.
8. Garzon AA, Gourin A. Surgical Management of Massive Hemoptysis, a 10 year experience. Ann Surg 1978; 187:267-271.
9. Rabkin J, Astafjen V, Gothman L et al. Transcatheter embolization in the management of pulmonary hemorrhage. Radiology 1987; 163:361-365.
10. Garzon AA. Massive hemoptysis: Surgical and tamponade therapy. In: Grillo HC, Austen WG, Wilkins EW, eds. Current Therapy in Cardiothoracic Surgery. St. Louis: C.V. Mosby Company, 1989:174-176.

Chest Wall Tumors

Fritz J. Baumgartner

Tumors of the chest wall fall into several classifications. They may be of bony or soft tissue origin. They may be benign or malignant and those that are malignant may be primary or metastatic. They may be lesions that invade the chest wall from the lung, pleura, mediastinum or breast.

Benign rib tumors include osteochondroma, chondroma, fibrous dysplasia and histiocytosis X. Malignant rib tumors include multiple myeloma, osteosarcoma, chondrosarcoma and Ewing's sarcoma. Osteochondroma is the most common benign bone neoplasm, constituting 50% of benign rib tumors. Multiple myelomas are the most common primary malignant rib neoplasm accounting for one-third of all malignant rib tumors, with chondrosarcoma being a close second. Benign soft tissue chest tumors include fibromas, lipomas, neurogenic tumors, hemangiomas and desmoid tumors.

Malignant fibrous histiocytoma (MFH) is the most common malignant soft tissue tumor. The tumor characteristically occurs in late adulthood and is rare in childhood. The disease often is a painless, slowly growing mass and is more common after chest wall irradiation. Malignant fibrous histiocytoma can spread widely along fascial planes or between muscle fibers accounting for a high recurrence rate after resection. MFH is unresponsive to radiation and chemotherapy and should be widely resected. Five year survival overall is approximately 40%.

Rhabdomyosarcoma is the second most common chest wall soft tissue malignancy. Wide resection followed by radiation and chemotherapy results in 5 year survival of about 70%. Liposarcoma and leiyomyosarcoma are other forms of soft tissue tumors. With regard to osteochondroma, the onset of pain in a previously asymptomatic tumor may indicate malignant degeneration. The neoplasm usually begins in childhood and continues to grow until skeletal maturity. Osteochondromas arise from the metastasis.

Chondromas constitute 15% of benign neoplasms of the rib cage. The differentiation between chondroma and chondrosarcoma is impossible on clinical and x-ray findings. Also, microscopically, the chondroma and a low grade chondrosarcoma may look identical; therefore, all chondromas must be considered malignant and should be treated by wide excision. Although this resection may appear to be excessive, the risk is negligible and may be life-saving.

Fibrous dysplasia presents, in most cases, as solitary lesions. However when multiple lesions are encountered (polyostotic fibrous dysplasia), then Albright syndrome may be present (multiple bone cyst, skin pigmentation and precocious sexual maturity in girls).

15

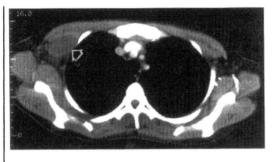

Fig. 15.1. Ewing's sarcoma in right anterolateral chest wall (arrow). The patient underwent resection of the tumor en bloc with the underlying chest wall and Marlex mesh reconstruction of the resulting defect.

Histiocytosis X is a disease involving the reticuloendothelial system and includes eosinophilic granuloma, Letterer-Siwe disease, and Hans-Schuller-Christian disease. Radiation therapy is frequently given. Steroid and chemotherapy may also be needed.

Myelomas involving the chest wall usually occur as a manifestation of systemic multiple myeloma. Solitary myeloma involving the rib is secondary only to solitary vertebral involvement. Multiple myeloma is rare under the age of 30. Most of the patients have abnormal electrophoresis and half have hypercalcemia and Bence-Jones protein in the urine.

Chondrosarcoma usually arises in the costochondral junction or the sternum. It is relatively uncommon under the age of 20. It most commonly presents as a slowly growing mass which has been painful for several months. All tumors arising in the costal cartilages should be considered malignant and should be treated by wide resection. The etiology of chondrosarcoma is unknown and may include malignant degeneration of chondromas, but may also be associated with trauma. Definitive diagnosis of chondrosarcoma can only be made pathologically; however even this is sometimes difficult because most chondrosarcomas are well differentiated. This is why wide resection of chondromas is necessary, since these may in fact be chondrosarcomas. The objective of the first operation should be resection wide enough to prevent local recurrence. This involves resection of a 4 cm margin of normal tissue on all sides. This results in cure in nearly all patients with a 10 year survival of 97%.

Ewing's sarcoma results in onionskin appearance of the surface of the bone on x-ray caused by elevation of the periosteum with subperiosteal new bone formation of multiple layers. Early spread to the lungs and other bones is common and occurs in about half of patients. Ewing's sarcoma is radiosensitive, so this modality is the treatment of choice. Five year survival is 50%. It may appear to be predominantly subcutaneous (Fig. 15.1).

Osteosarcoma is more malignant than chondrosarcoma and has a less favorable prognosis. Osteosarcoma usually occurs in teenagers and young adults who present with a rapidly enlarging, painful tumor. Alkaline phosphatase levels are frequently elevated. Calcification occurs at a right angle to the cortex producing a sunburst appearance on x-ray. Osteosarcoma treatment consists of wide resection of the tumor including the entire bone, rib or sternum and adjacent soft tissue.

15

The role of radiation and chemotherapy is controversial. The prognosis is poor with 5 year survival rates of 20%.

The management of suspected chest wall tumors include a careful history and physical examination followed by plain x-ray and CT scan. Tumors clinically suspected of being a primary neoplasm should be diagnosed by excisional rather than incisional or needle biopsy. However, if there is a history of primary neoplasm elsewhere and the chest wall mass may be a metastasis, then needle or incisional biopsy to establish diagnosis is acceptable. If a lesion is diagnosed by excisional biopsy as a malignant neoplasm, then further excision is indicated to include all of the involved bone and a 4 cm margin of normal tissue on all sides. This includes removal of the involved ribs and a rib above and below the neoplasm. For tumors of the sternum and manubrium, excision of the entire involved bone and corresponding costal arches bilaterally is indicated.

Metastases after they are diagnosed by biopsy do not require further surgical treatment. Radiation is the treatment of choice. Desmoid tumors tend to recur if inadequately excised and therefore should be treated with wide resection similar to primary malignant chest wall tumors. Bony defects should be reconstructed if large. Defects less than 5 cm in greatest diameter anywhere on the thorax are usually not necessarily reconstructed. Posterior defects less than 10 cm also do not require reconstruction because the scapula provides support. Larger defects elsewhere should be reconstructed. This can be performed with Prolene mesh or 2 mm Gortex soft tissue patch. Relative contraindications to patch implantation are an infected or necrotic field. Muscle flaps may be necessary to reconstruct the chest wall. This can be done with latissimus dorsi, pectoralis major, serratus anterior or rectus abdominus in descending order of preference.

SUGGESTED READING

1. Evans KG et al. Chest wall tumours. Can J Surg 1990; 33:229.
2. Fraley JH, Seyfer AE. Chest wall tumors: experience with 58 patients. Mil Med 1991; 156:413.
3. Graeber GB et al. Initial and long-term results in the management of primary chest wall neoplasms. Ann Thorac Surg 1982; 34:664.
4. Groff DB, Adkins PC. Chest wall tumors. Ann Thorac Surg 1967; 4:260.
5. King RM et al. Primary chest wall tumors: Factors affecting survival. Ann Thorac Surg 1986; 41:597.
6. Pairolero PC, Arnold PG. Chest wall tumors: Experience with 100 consecutive patients. J Thorac Cardiovasc Surg 1985; 90:367.
7. Ryan MB, McMurtrey MJ, Roth JA. Current management of chest wall tumors. Surg Clin N Am 1989; 1061.
8. Stelzer P, Gray WA Jr. Tumors of the chest wall. Surg Clin N Am 1980; 93:523.

15

Mesothelioma

Fritz J. Baumgartner

Mesothelioma is a primary tumor of the pleura. Most pleural tumors are actually not primary but rather metastatic from other primary sites such as breast or lung. The few primary tumors that do arise from the pleura are usually malignant. There is a strong causal relationship between asbestos exposure and mesothelioma. Nonetheless, only 7% of people exposed to asbestos actually develop mesothelioma. On the other hand, 50% of patients who have mesothelioma have history of exposure to asbestos. There is characteristically a long period of time between exposure and development of the tumor. The increased risk occurs 20 years after the first exposure and continues to rise for many years thereafter.

There are two types of mesothelioma—localized and diffuse. Diffuse mesotheliomas are nearly always malignant, however localized mesotheliomas may be either benign or malignant. About 30% of localized mesotheliomas are malignant and 70% benign. Benign pleural mesotheliomas usually arise from the visceral pleura on a stalk and project into the pleural space, although sessile attachment to the pleura may also occur. The tumor may also arise from the parietal pleura. Patients with localized malignant lesions usually have symptoms compared to patients with localized benign lesions; these include cough, pain, fever, and shortness of breath.

Mesothelioma may produce a bloody pleural effusion. In this case, complete resection of the lesion may still be possible even if it is associated with a bloody effusion. This is an important distinction between a mesothelioma and an intrathoracic lung neoplasm associated with a bloody pleural effusion. For a lung tumor, a pleural effusion which is bloody usually means an inoperable T4 lesion unlike mesothelioma.

Solitary mesotheliomas are usually asymptomatic. Benign localized mesothelioma is also accompanied by hypertrophic pulmonary osteoarthropathy. A large tumor may be associated with severe hypoglycemia as well. Hypertrophic pulmonary osteoarthropathy occurs in about half of patients with mesothelioma, compared with 5% of patient with bronchogenic carcinoma.

Clubbing of the fingers and toes may also occur with mesothelioma and is different from hypertrophic pulmonary osteoarthropathy. Clubbing results from new periosteal growth with lymphocytic infiltration of the nail beds. Whereas lung and pleural tumors may cause clubbing, benign causes include cyanotic heart disease, endocarditis, AV fistula, lung infection, COPD and sarcoid.

Treatment of localized benign mesothelioma is surgical resection, with good results. Treatment of the localized malignant variety is wide local excision, including adjacent chest wall if arising from parietal pleura.

Cardiothoracic Surgery, Third Edition, edited by Fritz J. Baumgartner. ©2004 Landes Bioscience.

In diffuse malignant mesothelioma, any portion of the pleura may be involved. It typically appears as sheets of tumor. Hematogenous spread to distant organs occurs in half. The symptoms include dyspnea, weight loss, cough and pain which is often severe. A pleural effusion is present in most of these patients. Diagnosis is achieved with needle pleural biopsy and thoracentesis. The treatment of diffuse mesothelioma is nearly always only palliative. However, radiation and chemotherapy have only been inconsistent in achieving this goal. Surgery in this disease is controversial. Seeding is frequent, and tumor may actually grow out of the thoracotomy incision used to treat the tumor. Some say surgery should be limited to achieving a diagnosis. Others advocate thoracotomy with pleurectomy whereas others perform a radical extrapleural pneumonectomy to remove all the pleura and lung on the affected side. Although 2 year survival after the procedure is improved with up to one-third surviving, 5 year survival is dismal.

SUGGESTED READING

1.   Rusch VW et al. Pleurectomy/decortication and adjuvant therapy for malignant mesothelioma. Chest 1993; 103:382S.
2.   Rusch VW, Venkatraman E. The importance of surgical staging in the treatment of malignant pleural mesothelioma. J Thorac Cardiovasc Surg 1996; 111:815.
3.   Sugarbaker DJ et al. Extrapleural pneumonectomy in the multimodality therapy of malignant pleural mesothelioma. Results in 120 consecutive patients. Ann Surg 1996; 224:288.

Esophagus

Fritz J. Baumgartner

ANATOMY

The esophagus is a continuation of the pharynx and ends at the cardia of the stomach. The transition from pharynx to esophagus is at the lower border at the 6th cervical vertebra which corresponds to the cricoid cartilage anteriorly. The esophagus deviates to the left in the lower portion of the neck, then returns to the midline in the mid-portion of the thorax near the bifurcation of the trachea. In the lower portion of the thorax the esophagus again deviates to the left to pass the diaphragmatic hiatus. Hence, the surgical maneuver of a right thoracotomy for a mid-esophageal lesion and a left thoracotomy for a lower esophageal lesion. The optimal exposure of the cervical esophagus is in the left neck not only because the esophagus deviates to the left in the neck, but also because the right recurrent laryngeal nerve is further away from the esophagus than the left recurrent laryngeal nerve. Hence, encircling the esophagus with the gloved finger is less likely to include the right recurrent nerve when dissecting from the left neck than it is to include the left recurrent nerve when dissecting from the right neck.

The average distance from the incisors to the cardia of the stomach, i.e., the distance for a complete esophagoscopy, is 40 cm. The esophagus itself is 25 cm long. The cricopharyngeus muscle (i.e., the beginning of the esophagus) is 15 cm from the incisors. The carina is at 25 cm from the incisors and lower esophageal sphincter is 40 cm from the incisors. These three numbers are important for determining anatomic location of tumors on esophagoscopy and their relationship to other intrathoracic structures. This is shown in Figure 17.1.

The fascial relations in the neck are clinically important. These are shown in Figure 17.2. The deep cervical fascia consists of four separate compartments or fascias. The pretracheal fascia surrounds the thyroid, trachea and the esophagus. Superiorly, this attaches to the thyroid and cricoid cartilage. Inferiorly, it extends to the pericardium and laterally blends with the carotid sheath. The investing cervical layer completely surrounds the neck and encloses the sternocleidomastoid and trapezius. Above, it splits to enclose the submandibular gland and parotid, and proceeds then to the base of the skull. The prevertebral fascia encloses the vertebral bodies and prevertebral muscles and extends superiorly to the base of the skull and inferiorly into the thorax. The space between the pharynx and the prevertebral fascia is the retropharyngeal space. It is through this space that infections arising in the deep cervical fascia, for example from dental abscesses, may

Cardiothoracic Surgery, Third Edition, edited by Fritz J. Baumgartner. ©2004 Landes Bioscience.

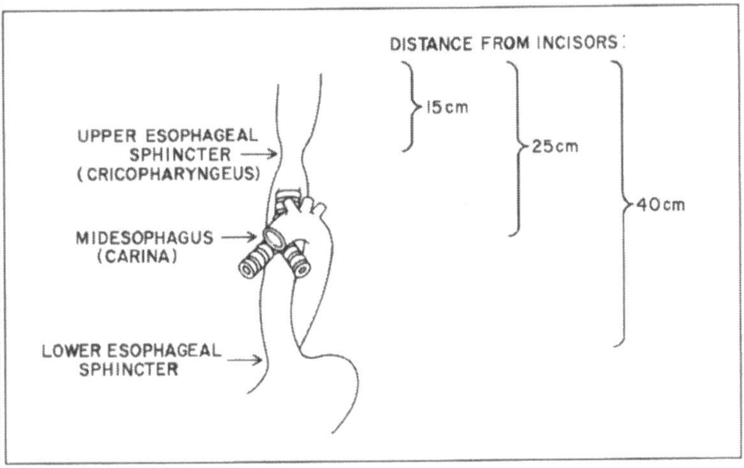

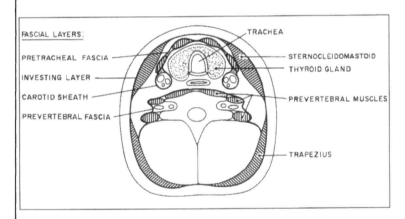

Fig. 17.1. Endoscopic esophageal distances (measured from incisors).

Fig. 17.2. Anatomy of the deep cervical fascia.

proceed further inferiorly to develop mediastinal abscesses in the posterior and middle mediastinum, and may even cause pericardial effusion.

Further down in the chest, the esophagus moves to the midline, then to the right. There is a natural narrowing of the esophagus behind the carina. There is another natural narrowing of the esophagus at the cricopharyngeus muscle, and another natural narrowing at the lower esophageal sphincter. It is at these three narrowings where perforations of the esophagus are more likely to occur after instrumentation.

The arterial blood supply of the esophagus is multiple. The upper portion of the esophagus receives blood from the inferior thyroid artery. The mid-portion of the esophagus receives arterial blood from bronchial arteries, and the lower most portion of the esophagus receives blood from the left gastric and inferior phrenic arteries. There is an extensive intramural vascular network in the muscular and submucosal layers, and as a consequence the esophagus can be mobilized from stomach to the aortic arch without fear of devascularization and ischemic necrosis. If the patient had a previous thyroidectomy with ligation of the inferior thyroid arteries proximal to the origin of the esophageal branches, care should be taken in the extent of the esophageal mobilization. This is one reason for taking the inferior thyroid artery close to the thyroid gland during a thyroidectomy.

Venous drainage of the esophagus is via a submucosal venous plexus which enters a peri-esophageal venous plexus. In the cervical region, the veins drain into the inferior thyroid vein; in the thoracic region they drain into the bronchial azygous or hemiazygous veins; and in the abdominal region into the coronary (i.e., left gastric) veins.

The esophagus contains squamous epithelium in the upper two-thirds of the esophagus and columnar epithelium in the lower third. There is a submucosal layer and a muscular layer with an inner circular coat and a longitudinal outer coat.

The lymphatic drainage from the cervical esophagus is into the internal jugular nodes and supraclavicular nodes. Drainage may occur into the paratracheal nodes. Drainage from the thoracic portion of the esophagus is into the para-esophageal nodes and paratracheal nodes, as well as inferiorly to the abdominal nodes. This area also drains superiorly into the supraclavicular nodes. The lower thoracic esophagus drains into the para-esophageal and infradiaphragmatic nodes, specifically the left gastric node, celiac nodes and hepatic nodes. There is some evidence that upper esophageal carcinoma is more likely to spread to the infradiaphragmatic nodes and that lower esophageal carcinoma is likely to spread to the supraclavicular nodes; i.e., the general trend of esophageal carcinoma is to go inferiorly rather than superiorly. The likelihood of nodal spread of the lower esophagus to go to the intra-abdominal nodes is much higher than the chance of upper esophageal carcinoma going to the supraclavicular nodes.

ESOPHAGEAL NEOPLASIA

BENIGN TUMOR

Leiomyoma, an intramural muscular tumor, is the most frequent benign neoplasm of the esophagus. It accounts for 75% of all benign esophageal neoplasms. Intraluminal tumors are very rare, and the majority of these lesions originate in the upper third of the esophagus and are generally benign polypoid tumors consisting of fibrous adipose and myxomatous material. They can be considered as fibrolipomas.

The majority of leiomyomas occur in the lower two-thirds of the esophagus where smooth muscle predominates. The leiomyomas generally have a fairly classic

17

appearance on barium swallow and are quite smooth. On esophagoscopy, a bulge may be seen into the lumen of the esophagus, however, the mucosa of the esophagus is intact. The diagnosis should not be attempted by biopsies through the intact mucosa. This is unnecessary and undesirable since violation of the mucosa will make later enucleation of the mass much more difficult.

There is some justification for not resecting a benign leiomyoma; however usually there is a need to rule out malignancy. A thoracotomy is generally done and a myotomy performed with enucleation of the leiomyoma with preservation of the mucosa. A left thoracotomy at the sixth or seventh interspace is performed for low lesions while a right thoracotomy is preferable for more proximal lesions. If there is massive esophageal involvement, then conservative resection of the involved esophagus and end-to-end anastomosis is preferred. A cervical approach may be necessary for cervical esophageal leiomyomas.

ESOPHAGEAL CARCINOMA

This is a difficult, confusing and frustrating subject. By the time most patients with esophageal carcinoma present, they are unresectable, although esophageal resection for cure in early esophageal carcinoma is possible and should be done. There are many alternatives to management and we will present our philosophy and approach, and then present the alternatives.

Tumors should be divided into cervical esophageal carcinoma, mid-thoracic esophageal carcinoma and lower esophageal carcinoma. The management of these lesions differs, as does their prognosis. The fundamental questions to ask in any tumor of the esophagus is: Is the lesion resectable and is the patient a surgical candidate who can survive a major operation, (i.e., these are the same questions to ask in lung carcinoma or in fact carcinoma in any portion of the body).

To evaluate whether the lesion is resectable, clinical staging of esophageal carcinoma is necessary. This is according to the TNM system. A T1 lesion does not extend through to the muscle wall; a T2 lesion does. A T3 lesion extends outside the muscle wall. Lymph nodes are N1 (unilateral), N2 (bilateral) or N3 (extensive and multiple). The M system is either, M0 or M1, i.e., visceral metastases or involvement of distant nodes (i.e., cervical or celiac).

Stage I is a T1 lesion; Stage II is a T2 lesion; Stage III is a T3 lesion or involvement of lymph nodes without metastatic lesions. These are generally considered inoperable. The management of stage I and stage II esophageal carcinoma is controversial, and the role of preoperative radiation therapy is likewise controversial. There is some evidence that radiation therapy prior to resection of esophageal carcinoma improves disease free survival, although an overall improvement in survival has yet to be conclusively demonstrated. The gastric nodes should be considered as N1 lymph nodes, rather than distant node metastases.

The other important issue is the patient as a surgical candidate. These patients generally are malnourished because of difficulty eating and may require a period of hyperalimentation to improve their nutritional status to make them a proper surgical candidate. It is important to evaluate their lung function so that it will be established that they can survive an operation without postoperative total ventilator dependence and that they can be weaned from a ventilator. Other co-morbid conditions need to be evaluated since these people generally are smokers who have concomitant lung carcinoma, emphysema or bronchitis, or may have other vascular complications related to their smoking; i.e., peripheral vascular disease including carotid, aortic and lower extremity vasculopathy, as well as coronary artery disease.

MALIGNANT TUMORS OF THE LOWER THIRD
OF THE ESOPHAGUS

After it is established that the patient is a surgical candidate and that the tumor is resectable, there are several surgical procedures that can be successfully used. It is important to note that tumors of the gastric cardia can be included in the category of malignant tumors of the lower third of the esophagus. These tumors are usually adenocarcinomas, whereas most tumors of the esophagus are squamous cell carcinomas.

Resection of malignant tumors of the lower esophagus consists of esophagogastrectomy including removal of the esophagus and proximal stomach with margins of at least 5 cm at either end, as well as resection in continuity of para-esophageal and left gastric and celiac axis lymph nodes. The extent of proximal esophageal resection is the most controversial, i.e., should only a 5 cm margin be achieved with an anastomosis in the chest or should the entire esophagus be removed? Our approach is to perform a total esophagectomy for several reasons.

First, removing the entire esophagus permits resection of any further possible source of malignancy in the proximal esophagus. It has been found that there is some frequency in which proximal esophageal carcinoma occurs simultaneous with distal esophageal carcinoma, and removal of the entire esophagus obviates this. Additionally, total esophagectomy performed via the transhiatal approach obviates the need for a thoracotomy. An Ivor-Lewis resection with gastroesophageal anastomosis, which will be described later, is a useful procedure and may be done depending on the surgeon's preference.

The transhiatal approach to esophagectomy is as follows: The patient is placed in the supine position and an upper midline abdominal excision is performed. An exploration and assessment of resectability is done. This is performed by taking down the greater omentum and the left gastroepiploic and short gastric vessels to mobilize the greater curvature of the stomach. The lesser sac is then entered and the left gastric vessel taken via the lesser sac. The esophagus is mobilized at the hiatus and a Penrose drain placed about it. The hand is placed up the hiatus, taking as many perforating vessels as possible with clips. One then mobilizes up into the mediastinum going behind the heart and carina and watching for hemody-

17

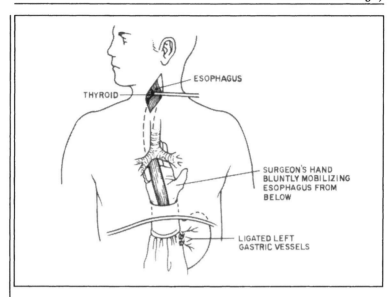

Fig. 17.3. Blunt transhiatal esophagectomy. The stomach has been mobilized in the abdomen and the esophagus has been mobilized in the neck. The surgeon's hand bluntly frees the esophagus within the mediastinum. If the esophagus is felt to be adherent to mediastinal structures, direct exposure is obtained through a thoracotomy or the procedure is terminated.

namic stability, so as not to compress the heart too much with the hand (Fig. 17.3). With the esophagus completely immobilized high into the mediastinum, an extensive Kocher maneuver is done, as well as a pyloromyotomy, which may be done in one of several ways. One is to incise the serosa and muscular coat of the pylorus, leaving the mucosa intact. Another way is to finger-fracture the pylorus and simply fracture the mucosa and muscular layers. The left gastric and left gastroepiploic vessels are taken. Hence, the entire blood supply of the stomach comes from the right gastric and right gastroepiploic vessels and these should be left intact.

While the abdominal mobilization is being performed, a second team of surgeons operates via a left neck approach to mobilize the cervical esophagus. An incision is made parallel to the sternocleidomastoid retracting the muscle and carotid sheath laterally and the thyroid gland medially. A finger is placed on the anterior portion of the vertebral body and a space dissected retroesophageally. The gloved finger is used to surround the esophagus with special care on the right side to hug the surface of the esophagus to avoid the right recurrent nerve. This is an important point. Most surgeons like to dissect the esophagus in the left neck not only because the esophagus lies more on the left than on the right in the neck, but also because there is less chance of injury to the recurrent nerve on the right side when the gloved finger is used to go around the esophagus from the left side.

This, however, is not an absolute surgical principle, and some surgeons still do the esophageal dissection from the right neck.

The esophagus is then mobilized deep into the superiormost portion of the mediastinum. This is facilitated by transsection of the strap muscles including the omohyoid. The esophagus is then hemi-transsected. The vein-stripper is passed into the stomach from below through a hole in the lesser curve of the stomach. The vein-stripper is then passed up out of the esophageal hole the neck, and the esophagus is transsected completely and stripped down from above into the abdomen. The GIA staple gun is then used across the angle of His down to the lesser curve to resect this portion of the stomach, taking care not to injure the right gastric artery.

The esophagogastrectomy specimen is then sent to pathology and the margins checked on frozen section. A moist, plastic bag is placed on the stomach remnant and suction applied to the stomach bag going through the mediastinum. The entire apparatus is smoothly dragged up the mediastinum into the neck. The plastic bag allows for a smooth placement of the stomach into the neck. The two layer anastomosis is then performed between the antrum of the stomach to the esophagus. A Penrose drain is placed in the neck and feeding jejunostomy performed.

An additional advantage of this approach is that there is no intrathoracic anastomosis which may leak and cause sepsis. If a leak occurs in the cervical esophagogastrostomy portion, this is a much less lethal condition and can be easily drained via the Penrose neck drain.

MALIGNANT TUMORS OF THE MIDDLE THIRD
OF THE ESOPHAGUS

Lesions in the mid-esophagus, constitute a much more difficult situation because of the possibility of local invasion into major mediastinal structures, in particular into the trachea and mainstem bronchi, as well as hilar vessels. First one must try to assess resectability in terms of the lesion; this is initially done by examination of the patient followed by CT scan to assess the proximity of the lesion to other major mediastinal structures. It is very important also to do bronchoscopy in these patients to assess the airway. However if there is still a question of encasement of the major mediastinal structures and the patient is still considered a candidate for surgery, the patient should be positioned for a right posterolateral thoracotomy and a 5th and 6th intercostal space incision made. The entire esophagus is mobilized and the tumor dissected free of surrounding structures after ensuring that this is indeed possible.

After the esophagus is entirely mobilized, but without transsecting it, the patient's chest is closed with a chest tube inserted for drainage. The patient then lays supine and a standard laparotomy and cervical dissection are performed for a transhiatal esophagectomy. Figure 17.4 illustrates a patient with esophageal cancer managed in this manner.

17

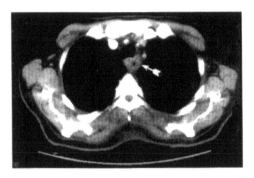

Fig. 17.4. CT scan of a patient with adenocarcinoma at the junction of the proximal and mid-portions of the esophagus (arrow). The lesion is seen abutting the membranous portion of the trachea. Bronchoscopy did not reveal tracheal invasion. A right thoracotomy was done to mobilize the tumor and esophagus. The patient was repositioned supine and a transhiatal esophagectomy with gastric pull-up into the neck was performed.

MALIGNANT TUMORS OF THE PROXIMAL THIRD OF THE ESOPHAGUS

For cervical esophageal carcinoma, radiation is the treatment of choice, not surgery. If radiation therapy fails with erosion of the tumor, for example into the larynx, then a laryngopharyngoesophagectomy is the operation of choice.

Squamous cell carcinoma of the esophagus is least common in this surgical region. Its management is difficult since it may involve larynx and trachea. It may metastasize to cervical lymph nodes as well. The evaluation in a patient with carcinoma of the cervical esophagus includes careful examination of the neck as well as barium x-ray films and CT scans of the neck and mediastinum, as well as the upper abdomen. Careful endoscopy of the larynx, trachea and esophagus are essential. Particular attention must be paid to laryngeal function and the distance between the cricopharyngeus and the tumor.

The likelihood of obtaining prolonged palliation in patients with extensive disease is minimal. In these advanced cases, radiation with or without chemotherapy may be used but without much chance of cure. Placement of an endoesophageal prosthesis is usually impossible at this level. More limited lesions are categorized into those involving the larynx and those that do not. For more limited lesions, if the carcinoma does invade the cricoid or larynx, it is not possible to salvage a functional larynx, and the resection must include larynx and enough trachea to provide a margin below the tumor, as well as removal of the entire cervical esophagus. A mediastinal tracheostomy is required. In this procedure, a collar incision provides access to both sides of the esophagus. A flap is raised and the extent of involvement is completely assessed before definitive resection is performed. The isthmus of the thyroid is transected and the trachea is dissected to a point low enough to define a margin of normal tissue. Intubation is carried out across the operative field into the transected distal trachea. The esophagus is transected in the mediastinum. If total esophagectomy is undertaken, the dissection can be carried down to the carina from the cervical approach, and the transhiatal esophagectomy completed from the abdominal approach.

The specimen with the esophageal tumor is elevated and the lateral attachments of the larynx on either side are divided. The pharynx is entered through an incision above the hyoid bone and carried laterally. The epiglottis is removed at the hyoid bone. If total esophagectomy is performed, it is simplest to mobilize the stomach for replacement and bring it up for a direct pharyngeal-gastric anastomosis. If metastatic lymph nodes are found in the neck, cervical lymph node dissection may be done at the same time.

For cervical lesions not involving the larynx, initial dissection and exploration of the lesion are more difficult. The esophagus is encircled below the lesion at a reasonable point that will provide a satisfactory distal margin. This is done through a collar incision, since then each side of the esophagus can be dissected. In the event a laryngectomy is needed, this can be done through the collar incision. Again, a total esophagectomy is performed and a pharyngeal-to-stomach anastomosis is performed in the cervical region.

IVOR-LEWIS TECHNIQUE

All of the above treatment modalities for tumors in the cervical, mid-thoracic and lower esophageal regions are acceptable surgical positions. These are not the only choices and many other procedures exist. One popular method of treatment, for example, for lower esophageal carcinoma is the Ivor-Lewis procedure. In this procedure, the stomach is brought up into the right chest and a thoracic esophageal-to-stomach anastomosis is performed. This leaves the proximal esophagus intact. The disadvantage of this is that there still may be multicentricity of tumor which could involve the proximal esophagus. This is partially obviated by the fact that most lower esophageal carcinomas are adenocarcinoma in histology and are therefore probably of gastric origin and do not have the same incidence of multicentricity seen with squamous cell carcinoma. The second major reason that an Ivor-Lewis procedure may be more fraught with complications than a transhiatal esophagectomy is the fact that a thoracic anastomotic leak is a much more disastrous situation than a cervical esophageal anastomotic leak. The intrathoracic leak can lead to empyema and can rapidly progress to death, whereas a cervical esophageal anastomotic leak can be managed simply with a drain for weeks if need be.

An anastomosis may be technically perfect; however if the region becomes devascularized for any reason, the chance of healing properly is diminished and a leak may be inevitable. Therefore, it is not a trivial problem even in the best of hands.

In the Ivor-Lewis procedure, there again are several options available. One is to have the patient laying at a 45 degree angle and perform a thoraco-abdominal incision through the seventh intercostal space, mobilize the stomach, mobilize the esophagus, transect the esophagus and bring the stomach up into the chest to perform the intrathoracic esophagogastrostomy. A drainage procedure is necessary and this usually involves a pyloroplasty, either as a finger fracture pyloroplasty or a myotomy incision in the seromuscular layer down to the mucosa.

The other option is to perform separate laparotomy and thoracotomy. In this case, the patient is supine and an upper midline incision is made and the stomach mobilized and brought up through the diaphragm. The patient is placed in the lateral decubitus position with the right side up and the thoracotomy performed. The stomach is further brought up into the chest, the esophagus mobilized; and an esophagogastric anastomosis performed.

COLONIC/JEJUNAL ESOPHAGEAL REPLACEMENT

Once it is established that esophageal resection is indicated, an organ for esophageal replacement becomes necessary. Although the stomach is the usual organ for esophageal replacement, the colon or jejunum also can be used. The stomach, in general, is sturdier and has more reliable blood supply. Care must be taken that the replacement organ is usable and that it is pathologically free of disease. Endoscopy, arteriography and contrast radiography are useful to determine this. Not all of these are necessary, for example when using the stomach as the conduit organ. However, endoscopy should always be performed to rule out ulceration or malignancies.

For colonic bypass, colonoscopy should be performed preoperatively, as well as angiography to evaluate the colonic vessels. A barium enema should be performed last (since it obscures the angiogram dye). Bowel preparation preoperatively is mandatory.

The colon bypass will now be explained in detail. The patient is supine and an upper midline incision performed. The omentum is detached from the transverse colon. The peritoneal reflection on the left is taken down. The middle colic artery and vein are ligated near the origin at the superior mesenteric vessels. The transverse colon is transected proximally using the GIA stapler. The mesentery of the sigmoid is transilluminated to identify the left colic artery, which will be included with the replaced colonic segment. The colon segment to be used is transsected at the sigmoid region and this segment is now attached only to its supplying artery and vein. It is passed posterior to the stomach through a rent in the gastrohepatic omentum and this allows the most direct alignment of the artery and vein. The esophagus is bluntly removed through the abdominal approach as well as the cervical approach as described previously. The esophagus is removed and the colonic segment brought up transhiatally into the left neck where an esophagocolonic anastomosis is performed in the neck and distally a colon-to-gastric anastomosis is performed. A pyloroplasty is performed to prevent gastric retention. A colocolotomy is then performed to anastomose the proximal transsected colon to the distal transsected colon to re-establish gastrointestinal continuity.

Another organ that is much more rarely used is a free jejunal graft. In this situation, a segment of jejunum is dissected with its vascular pedicle and bowel continuity is re-established with an end-to-end anastomosis. The free jejunal segment is then brought up into the neck and the superior thyroid lingual or facial arteries are anastomosed to the jejunal artery and vein. In addition, the transverse cervical artery at the base of the neck may be used. The free jejunal graft is best used for a localized resection. For example, in the case of cervical esophageal resection where only this portion of the esophagus is resected, an anastomosis can be performed, from the pharynx to the jejunum and then another anastomosis at the base of the neck.

UNRESECTABLE ESOPHAGEAL CANCER

This can be a terrible problem since it is a lethal condition in which the patients typically die miserably, aspirating on their own secretions unless some intervention is done. There are several treatment options available. One is opening a lumen through the tumor. This can be done with either esophageal stents (i.e., esophageal intubation) or by laser obliteration of the tumor stenosing the esophagus. Another is a bypass procedure in which the stomach is passed substernally into more proximal esophagus to bypass the tumor. This frequently results in major complications, usually because of the preoperative morbid status of the patients. In patients with obstructive esophageal neoplasms, palliation of the esophagus is the primary treatment goal. Radiation may also be effective in relieving the esophageal stenosis to some extent.

Each modality has its advantages and disadvantages. Advantages of esophageal intubation include that it can be done either under local or general anesthesia. There are a wide variety of diameters and lengths of esophageal tubes which can be tailored to fit the patient's needs. These disadvantages include perforation, slippage of the tube, and reflux of gastric contents with aspiration (this is particularly true if the tube passes across the gastroesophageal junction).

The typical patients who require intubation of the esophagus or laser obliteration of esophageal carcinoma are those with medical contraindications to resection, those who have local invasion of major mediastinal structures preventing resection, those with malignant tracheal esophageal fistulas, or those who are treated primarily with radiation therapy and then have recurrent esophageal obstruction. These are all, in general, contraindications to resection.

HIATAL HERNIA AND GASTROESOPHAGEAL REFLUX

The overall management of patients who present with dysphagia or heartburn must include, of course, a cardiovascular history and examination since the suspicion of ischemic cardiac disease or other myocardial dysfunction must always be in mind.

The esophageal problems that may result in dysphagia or heartburn can include carcinoma, gastroesophageal reflux or motility disorders (including achalasia or diffuse esophageal spasm). The preoperative evaluation of patients who present with suspected esophageal motility disease or gastroesophageal reflux should always include esophagoscopy (with biopsy if necessary); a 24-hour pH probe study; and manometry. An upper GI study may be helpful on occasion, although it is redundant. If the above three studies are done well, they will give all the information an upper GI study can provide and much more.

The 24-hour pH probe monitor determines with high sensitivity and specificity the presence of gastroesophageal reflux. This is much more difficult to evaluate using the upper GI barium swallow. An upper GI barium swallow which reveals a hiatal hernia is not at all synonymous with gastroesophageal reflux. Not all patients who have hiatal hernias have gastroesophageal reflux. In fact, only about 10% of patients with hiatal hernias have clinical gastroesophageal reflux. Also it should be noted that the barium swallow is not a specific test either. Reflux of barium into the esophagus can be observed in 25% of normal subjects and the inability to demonstrate reflux in an upright position does not exclude the presence of GE reflux.

The 24-hour esophageal pH monitoring test is performed by placing a pH electrode 5 cm above the upper border of the lower esophageal sphincter as measured manometrically. A normal diet is given with the exception that food or liquid with a pH less than 5 or greater than 7 is excluded. At the completion of the monitored period, the probe is advanced into the stomach to document the pH of the gastric contents. Patients are instructed to record symptoms over the 24 hours and this is then reviewed on the pH probe continuous strip. Acid reflux is identified whenever the pH in the esophagus drops to 4 or below. The 24-hour esophageal pH monitoring is the most accurate way to measure gastroesophageal reflux.

In the work-up of suspected gastroesophageal reflux with symptoms of heartburn and/or dysphagia, a careful history and physical examination are done followed by esophagoscopy, manometry and 24-hour pH testing. If these reveal esophageal reflux, medical management is indicated for a period of 6 weeks until the next evaluation.

Medical management includes the following: H2 blockers, antacids, and abstinence from alcohol and smoking (i.e., these are all the same medical management maneuvers used for peptic ulcer disease). In addition, the patient should sleep in an upright position, i.e., by placing shock blocks at the head of the bed. If the patient's symptoms have not resolved in 6 weeks, surgery may be indicated. This may be of several different types, the most commonly being a Nissen fundoplication.

Indications for performing surgery for gastroesophageal reflux includes the following: persistent esophagitis; bleeding; ulceration; severe dysplasia found on endoscopic biopsy; stricture; or recurrent aspiration. Although controversial at the present time, the presence of Barrett's esophagus does not mandate prophylactic anti-reflux procedure but should be followed closely.

PROCEDURES

The operations available for anti-reflux include the Nissen fundoplication, the transthoracic Belsey Mark IV procedure, the Hill fundoplication, or an intrathoracic Nissen fundoplication. Advances in laparoscopic technique have permitted

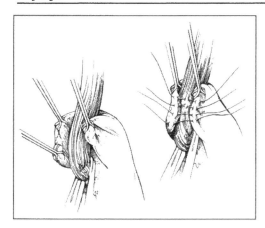

Fig. 17.5. Nissen fundoplication procedure. The fundus of the stomach is wrapped 360° about the esophagus. Reprinted with permission from: McFadden DW, Zinner MJ. Benign disease of the esophagus. In: Ritchie WP Jr, Steele G Jr, Dean RH, eds. General Surgery. Philadelphia: J.B. Lippincott Co, 1995:80.

endoscopic Nissen fundoplication as well.

The standard abdominal Nissen fundoplication is performed with the patient placed supine, and a standard upper midline incision is made. The greater curvature of the stomach is mobilized taking the short gastric arteries. A size 40 Maloney dilator is passed through the gastroesophageal junction from above and mobilized distal esophagus and proximal stomach is wrapped using the mobilized fundus of the stomach in a 360° fashion (Fig. 17.5). Interrupted sutures are then passed through the seromuscular coat of the stomach through the muscle area of the esophagus, then back out through the other side of the stomach. This is done for about a 2-3 cm length. It should be sufficiently loose that the gloved finger is able to be passed up through the repair between the fundus and the esophagus. Some people perform a tacking maneuver of the fundoplication down to the pre-aortic fascia, i.e., modified Hill procedure. However this is not necessary. The hiatal hernia defect is repaired with interrupted sutures.

One of the complications of a Nissen fundoplication is the gas-bloat syndrome which may result from too tight of a wrap or injury to the vagus nerves. The treatment is pyloroplasty drainage procedure.

For the transthoracic Belsey Mark IV, a lateral thoracotomy is performed at the 7th or 8th intercostal space. After complete mobilization of the cardia of the stomach, the lower 4 cm of the esophagus is cleared of connective tissue. Mattress sutures are placed between the gastric fundus and the muscle layers of the esophagus (Fig. 17.6). After these mattress sutures are tied, a second row of mattress sutures is placed between the muscle coat of the esophagus, the gastric fundus and the crus of the diaphragm. Before these sutures are tied, crural sutures are placed to narrow the esophageal hiatus; this results in an effective anti-reflux mechanism. It should be noted that the wrap is only 270° rather than the 360° Nissen fundoplication.

The Hill procedure for gastroesophageal reflux is a conceptually simple operation. It is basically a crural plication added to a posterior gastropexy. The crura of the diaphragm is approximated over the distal esophagus. A posterior gastropexy

17

Fig. 17.6. Transthoracic Belsey
Mark IV: A double-telescoping
procedure of the esophagus into
the stomach as a 270° wrap.

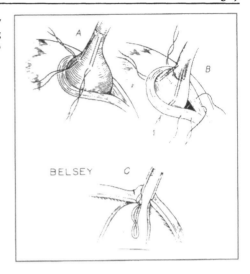

is performed from the cardia of the stomach down to the median arcuate ligament, i.e., pre-aortic fascia. This effectively stabilizes the lower esophageal sphincter, tightens the esophageal hiatus and limits gastroesophageal reflux.

There may be instances where the disease is so severe that the esophagus is
actually shrunk, the stomach is permanently fixed in the chest, and the stomach
cannot be reduced into the abdomen. In this case, an esophageal lengthening procedure may be necessary. This is commonly done with an addition of a Collis
procedure to either the Nissen or the Belsey Mark IV. In this case a staple gun is
used to cut down on the greater curvature side of the stomach, forming a neo-
esophagus. The fundus of the stomach can then be wrapped about the neo-esopha-
gus either in a 360° fashion for Nissen fundoplication either within the abdomen
or within the chest, or it may be used in a 270° fashion as in a Collis-Belsey Mark
IV done transthoracically.

What is the management of severe stricture of the esophagus? This is an indication for an anti-reflux procedure. After endoscopy and biopsy to ensure that
this is not a malignant stricture, dilation of the stricture should be performed
with progressively larger bougies followed by an anti-reflux procedure. It should
be noted that there may be instances of nondilatable stricture which may require
esophagectomy and replacement of either a part of or usually the entire esophagus. This is an ideal situation for a transhiatal esophagectomy with a cervical
esophagogastric anastomosis.

MOTILITY DISORDERS

Motility disorders are usually classified as either achalasia or diffuse esophageal spasm. Achalasia can be defined manometrically as failure of the lower esoph-

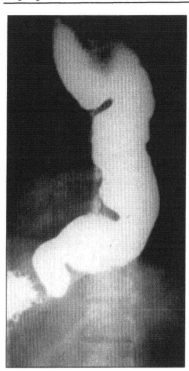

Fig. 17.7. Barium esophagram of a young woman with severe achalasia. She underwent a thoracoscopic esophagomyotomy with a curative outcome.

ageal sphincter to relax and lack of peristalsis of the proximal esophagus (Fig. 17.7). The treatment of achalasia is pneumatic dilation or a Heller esophagomyotomy. Diet modification and drugs are generally ineffective. Pneumatic dilation has an initial success rate in the range of 80%; however subsequently only less than 50% have long-term good results. Therefore in general we feel it should only be done for poor risk patients. Esophagomyotomy is probably the treatment of choice.

In a Heller esophagomyotomy, a lateral thoracotomy is performed in the 7th or 8th intercostal space. The muscular layer is taken down to the mucosa which is exposed and this is taken down 1 cm onto the stomach. One must dissect a plane between the mucosa and muscular layers lest the muscular layer reapproximates. The dissection is carried down to the junction of the esophagus with the stomach where the longitudinal submucosal venous plexus becomes the transverse venous plexus along the stomach. There is no need for an anti-reflux procedure since reflux only occurs in 3%, although some groups advocate a concomitant anti-reflux procedure with an esophagomyotomy. If recurrence of achalasia occurs after Heller esophagomyotomy, then another Heller esophagomyotomy is performed with more extensive dissection between the mucosal and muscular layers.

For diffuse esophageal spasm, pain is much more pronounced and the patients rarely complain primarily of dysphagia alone. An upper GI may reveal narrowing and pseudodiverticulosis. Initial medical management consists of nitrates and

17

calcium channel blockers; however these only work in about 50% of patients. Failure of these modalities requires a Heller esophagomyotomy but the proximal extent of the myotomy is longer compared to that for the treatment for achalasia. The level depends on preoperative esophageal motility studies which reveal the proximal extent of the esophageal spasm. If the lower esophageal sphincter is found to be normal manometrically, then the myotomy can stop short of the lower esophageal sphincter. It should be noted that surgical treatment for spasm using esophagomyotomy is less effective than for achalasia.

SUGGESTED READING

1.   Mathisen DJ, Grillo HC, Hilgenberg AD et al. Transthoracic esophagogastrectomy: A safe approach to carcinoma of the esophagus. Ann Thoracic Surg 1988; 45:137-143.

2.   Ellis FH Jr, Gibb SP, Watkins E Jr. Esophagogastrectomy: A safe, widely applicable, and expeditious form of palliation for patients with carcinoma of the esophagus and cardia. Ann Surg 1983; 198:531-540.

3.   Orringer MB. Transhiatal esophagectomy without thoracotomy for carcinoma of the thoracic esophagus. Ann Surg 1984; 200:282-288.

4.   Katlic M, Grillo HC. Carcinoma of the esophagus. In: Choi NC, Grillo HC, eds. Thoracic oncology. New York: Raven Press, 1983:279.

5.   Bare AE, Belsey RHR. The treatment of sliding hiatus hernia and reflux esophagitis by the Mark IV technique. Surgery 1967; 62:396-404.

6.   Skinner DB, Belsey RHR. Surgical management of esophageal reflux and hiatus hernia. J Thorac Cardiovasc Surg 1967; 53:33-54.

7.   Orringer MB. Surgical treatment of esophageal strictures resulting from gastroesophageal reflux. In: Stipa S, Belsey RHR, Moraldi A, eds. Medical and surgical problems of the esophagus. New York: Academic Press, 1987:165.

8.   Ellis FH Jr, Crozier RE, Watkins E Jr. Operation for esophageal achalasia: Results of esophagomyotomy without an antireflux operation. J Thorac Cardiovasc Surg 1984; 88:344-351.

9.   Pac M et al. Transhiatal vs. transthoracic esophagectomy for esophageal cancer. J Thorac Cardiovasc Surg 1993; 106:205.

10.   Moon MR et al. Transhiatal and transthoracic esophagectomy for adenocarcinoma of the esophagus. Arch Surg 1992; 127:951.

11.   Ellis FH et al. Esophagectomy for carcinoma of the esophagus and cardia: A comparison of findings and results after standard resection in three consecutive eight-year intervals with improved surgical criteria. J Thorac Cardiovasc Surg 1997; 113:836.

17

Mediastinal Masses

The mediastinum can be divided into three compartments—anterior, middle and posterior mediastinum (Fig. 18.1). The anterior mediastinum may be subcategorized into anterior and superior mediastinum. The superior portion of the anterior mediastinum lies above a plane extending from the sternal angle of Louis to the level of T4. The rest of the anterior mediastinum is in front of the anterior pericardium. The middle mediastinum is the space between the anterior portion of the pericardium back to the anterior portion of the vertebral bodies, and the posterior mediastinum is between the anterior portion of the vertebral body to the posterior portion of the vertebral body, i.e., to include the region of the paravertebral gutters. Tumors of the anterior mediastinum include thymoma, germ cell tumors (teratoma is one type), thyroid enlargement, and lymphoma. Middle mediastinal masses include pericardial cysts, bronchogenic cysts, lymphoma and mediastinal granulomas. Posterior mediastinal masses are predominantly tumors of neurogenic origin.

Evaluation of mediastinal tumors commences with the history and physical examination. The history is very important because a history of chills and night sweats may be more consistent with lymphoma rather than thymoma or germ cell tumor, for example. Physical examination, including a good systemic examination, is performed. Laboratories including alpha-fetoprotein and beta HCG for anterior mediastinal tumors, as well as catecholamine levels and urinary VMA levels are important for posterior mediastinal tumors.

There are three approaches for tissue assessment of mediastinal tumors: Percutaneous needle aspiration; mediastinoscopy; and anterior mediastinotomy (Chamberlain procedure). In general, needle aspiration biopsy of presumed thymomas is contraindicated because of the possibility of seeding the mediastinum or pleura. Mediastinoscopy is excellent for evaluating adenopathy in the mediastinum including diagnosis of a lymphoma. All mediastinal tumors require investigation by CT for evaluation of the lesion itself, as well to evaluate invasion of surrounding structures. A mediastinal tumor which is suspected of having a vascular component should be investigated with angiography.

ANTERIOR MEDIASTINUM

Anterior mediastinal tumors may consist of thyroid enlargement for the superior portion of the anterior mediastinum. This usually involves a thyroid goiter, but

Cardiothoracic Surgery, Third Edition, edited by Fritz J. Baumgartner. ©2004 Landes Bioscience.

Fig. 18.1. Mediastinal compartments.

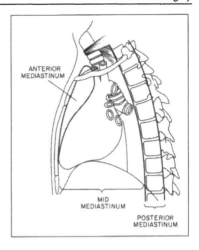

may involve a thyroid tumor.

Thymomas are the most common anterior mediastinal mass. They are usually benign but may be malignant. The interrelationship between thymomas and myasthenia gravis will be discussed subsequently. Suffice it to say that 10% of patients with myasthenia gravis have thymomas, whereas half of patients with thymomas have myasthenia gravis. The thymoma may be benign or malignant with that determination being made by the surgeon at the time of surgery, not by the pathologist since the histology will appear to be the same microscopically and the invasive features are the most reliable means to determine malignancy (Fig. 18.2).

Germ cell tumors can be either benign or malignant. Benign tumors include benign germ cell tumors and benign teratoma or dermoid cyst. The malignant germ cell tumors are much rarer than benign ones. The malignant germ cell tumors may be seminomanous or nonseminomanous. Patients with seminomanous malignant germ cell tumors must undergo ultrasound of the testicles even if physical examination is negative since there may be a nonpalpable focus of tumor still there with the mediastinal component being a metastasis rather than a primary tumor. Seminomanous tumors have low levels of beta HCG and alpha-fetoprotein unlike its nonseminomanous germ cell counterpart.

The management of seminomanous germ cell tumors when small and localized is surgical resection followed by radiation. This results in nearly 100% cure rate; 60% of these lesions are cured with radiation therapy alone (this is compared to 80% cure rate for radiation therapy for testicular seminomas). If there are distant metastases from the mediastinal primary or testicular primary, then chemotherapy should be utilized with radiation therapy to the primary. For locally advanced disease without metastases, the treatment is controversial but should probably involve radiation therapy, and for relapses chemotherapy.

Nonseminomanous malignant germ cell tumors include embryonal carcinoma, malignant teratoma, choriocarcinoma or endodermal sinus tumor (yoke sac tumor). These tumors generally have elevated beta HCG and alpha-fetoprotein. The

18

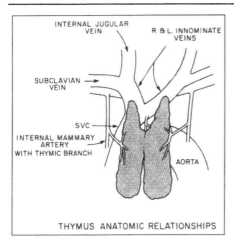

Fig. 18.2. Normal thymus anatomic relationships.

management of these tumors is chemotherapy (initially cisplatin, bleomycin and vinblastine regimen) and if there is a response with decrease in the beta HCG/ alpha fetoprotein levels, then resection is indicated. Generally, if there is no decrease in beta HCG and alpha-fetoprotein levels with chemotherapy, then this signifies a relatively aggressive tumor and resection is not indicated. The role of adjuvant radiotherapy is not settled at this point.

MYASTHENIA GRAVIS AND THYMOMA

Myasthenia gravis is a condition in which the patient develops progressive muscle weakness secondary to acetylcholine receptor antibodies. Medications used to treat myasthenia gravis include anti-cholinesterase medications which block the cholinesterase-mediated removal of the acetylcholine from its receptor. The classifications of myasthenia gravis include ocular and generalized. In the ocular group, there is involvement of ocular muscles only with ptosis and diplopia. This is a very mild form of myasthenia gravis with no mortality. The generalized type of myasthenia gravis may be mild or progress to severe disease. In the more severe forms, the response to drug therapy is poor and the prognosis is poor. Ninety percent of patients with myasthenia gravis have adult onset disease. It affects women twice as often as men. Ten percent of patients with myasthenia gravis have thymomas. One-third to one-half of patients with thymomas have myasthenia gravis.

Myasthenia crisis is an episode of severe, rapid progression of the disease with respiratory compromise. Patients with ocular symptoms alone or those whose conditions are stable and well-managed with medication are not generally candidates for thymectomy because their prognosis is generally good. On the other hand, patients in myasthenia crisis are also not candidates for emergency surgery, and thymectomy can be delayed until better medical control is obtained.

Myasthenia patients are not particular anesthetic risks, although no muscle

18

relaxant should be used during the procedure. For patients with severe my-asthenia gravis, several options for treatment exist, including medications and plasmapheresis which may be a useful preoperative adjunct by reducing circulating levels of acetylcholine receptor antibodies.

The reason that thymectomy is effective in myasthenia gravis is unknown. It has been shown that patients with a normal thymus who undergo thymectomy for myasthenia gravis do better in terms of decreased requirement for medication than those patients with myasthenia gravis who have thymomas resected. The reason for this is unknown. In general, it can be expected that in those patients with myasthenia gravis who undergo thymectomy, one-third will be cured of their disease; one-third will be somewhat improved; and one-third will have no benefit. A poorly understood paradoxical phenomenon is the rare de novo appearance of myasthenia gravis after resection of a thymoma, occurring in less than 5% of thymomas.

The thymus lies in the anterior mediastinum overlying the pericardium and great vessels. The arterial supply is predominantly from the internal mammary vessels, whereas the venous drainage is central into the left innominate vein over which it is draped. It is an H-shaped organ with two limbs going up into the neck and two limbs going down into the anterior mediastinum (Fig. 18.2).

Surgical approaches to the thymus include a transcervical approach, a limited median sternotomy and a full median sternotomy. In the cervical approach, a collar incision is made and the thymus is dissected out through this approach. As can be imagined, it is extremely difficult to resect the thymus in its entirety using this approach, and we would generally not recommend it. The limited median sternotomy approach allows better access to the upper mediastinum. In this approach, an upper midline incision is made and a saw is used to perform a median sternotomy down to the level of about the 3rd intercostal space, at which point the saw is turned transversely and the sternum transsected. A small retractor is then used to spread the transsected upper sternum, and thus a limited upper median sternotomy is performed. A full median sternotomy allows the best visualization of the anterior mediastinum.

The thymus is completely dissected free and one may need to go all the way down on either side of the pericardium towards the phrenic nerves for complete resection.

In the case of a thymoma, lack of an enveloping capsule or local invasion of adjacent tissues including pericardium, pleura, great vessels or phrenic nerve is a good indication of malignancy. Thymomas usually do not metastasize by hematogenous or lymphatic roots, but metastasize directly to pleural surfaces by seeding or by direct extension. The thymoma may recur locally if its capsule has been violated. If the tumor appears invasive, radical excision may be warranted. The phrenic nerve needs to be identified and the tumor may be cleared away from the phrenic nerve if it is not actively invading the nerve.

It is controversial whether or not a phrenic nerve should be taken. Generally, if the patient does not have myasthenia gravis, then sacrifice of one of the phrenic nerves is acceptable, since these patients will generally still be able to be extubated, if they have a normally functioning hemidiaphragm on the other side. If, however, they have myasthenia gravis, it is controversial whether or not resection of a phrenic nerve is acceptable. It appears logical that if it is being actively invaded, then it probably is not functional and its sacrifice may be warranted. Removal of invaded pericardium may be performed as well as removal of adherent mediastinal pleura and wedge excision of portions of lung that are invaded.

Preoperative radiation therapy may be considered if the thymoma is huge or if radiologic evaluation suggests vascular invasion. The dose is in the range of 4000 rads over 4 weeks followed by surgery in another 4 weeks.

MIDDLE MEDIASTINUM

The tumors of the middle mediastinum include lymphoma, pericardial cyst, bronchial cyst and mediastinal granuloma. These can frequently be identified on CT scan. A pericardial cyst may not need treatment at all other than biopsy for identification, and if this has a classic appearance on CT scan, even this may not be necessary.

Diagnosis of lymphoma usually by anterior mediastinotomy or cervical mediastinoscopy is required. Bronchial cyst usually originates from the mainstem bronchi and usually requires resection by thoracotomy to prevent recurrent infection. Mediastinal granulomas do not need treatment unless they are actually eroding into the tracheobronchial tree, such as is the case with broncholithiasis. This generally requires thoracotomy for removal of the broncholith or rarely rigid bronchoscopy if thoracotomy is contraindicated.

POSTERIOR MEDIASTINUM

The most common neurogenic tumor in the posterior mediastinum is a benign neurolemmoma; in children the most common neurogenic tumor is the malignant neuroblastoma. The histologic classification is predominantly that of nerve sheath tumors and ganglionic/paraganglionic tumors. The benign nerve sheath tumors is the neurolemmoma or neurofibroma; its malignant counterpart is the malignant schwannoma (Figs. 18.3, 18.4, 18.5). The autonomic tumors are the benign ganglioneuroma; its malignant counterpart is the ganglioneuroblastoma or the malignant neuroblastoma. Pheochromocytomas are also autonomic tumors and 90% of them are benign. CT scanning is an important part of the diagnostic armamentarium. Serum epinephrine and norepinephrine levels should be obtained as well as urinary VMA levels.

18

Fig. 18.3. Benign thymoma seen at the time of surgery. No evidence of gross invasion is seen.

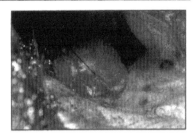

Fig. 18.4. Schwannoma (+) in posterior mediastinum invading the vertebral body. Resection of the tumor and vertebral body was performed followed by a spinal stabilization procedure.

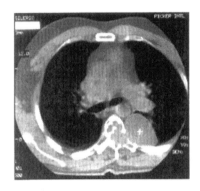

Fig. 18.5. Schwannoma in posterior mediastinum in a more unusual superiorly located position.

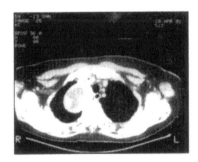

SUGGESTED READING

1. Bergh NP, Gatzinsky P, Larsson S et al. Tumors of the thymus and thymic region: I. Clinicopathologic Studies on thymomas. Ann Thorac Surg 1978; 25:91-98.
2. Wilkins EW Jr. Thymectomy. In: Cohn LH ed. Modern techniques in surgery (cardiac/thoracic surgery). Mount Kisco: Futura Publishing, 1981; 38:1-13.
3. Wychulis AR, Payne WS, Clagett OT et al. Surgical treatment of mediastinal tumors. A 40 year experience. J Thorac Cardiovasc Surg 1971; 62:379-392.
4. Marchevsky AM, Kaneko M. Surgical pathology of the mediastinum. New York: Raven Press, 1984.
5. Hsu CP et al. Thymic carcinoma. Ten years' experience in twenty patients. J Thorac Cardiovasc Surg 1994; 107:615

6. Namba T, Brunner NG, Grob D. Myasthenia gravis in patients with thymoma, with particular reference to onset after thymectomy. Medicine 1978; 57:411

7. Allen MS, Trastek VF, Pairolero PC. Benign germ cell tumors of the mediastinum. In: Shields TW, ed. General Thoracic Surgery. 5th ed. Lippincott Williams & Wilkins, Philadelphia, 2000:2275-88.

18

Thoracic Trauma

Fritz J. Baumgartner

The management of thoracic trauma is a complex field not only because of the major anatomic structures running through the chest, but also because management of thoracic trauma must be integrated with management of concomitant abdominal trauma, neurologic trauma and orthopedic trauma. Ignoring any of these aspects may be life-threatening for the patient.

The immediately life-threatening chest injuries that need to be identified by the physician in the initial assessment include the following:

1. airway obstruction
2. tension pneumothorax
3. open pneumothorax (i.e., sucking chest wound)
4. massive hemothorax
5. flail chest
6. cardiac tamponade

During the secondary survey, the identification and treatment of other life-threatening injuries can be achieved. These include:

1. aortic disruption
2. pulmonary contusion
3. tracheobronchial rupture
4. myocardial contusion
5. esophageal rupture
6. traumatic rupture of the diaphragm

Most injuries to the chest, both blunt and penetrating, do not require surgery. Only about 10% of chest injuries actually require operative management. Hypoxia is the most immediately threatening feature of chest injury and interventions directed towards managing chest trauma must insure adequate ventilation and oxygenation of the body.

The primary survey of life-threatening chest trauma includes an assessment of the adequacy of the ventilation, and if this is inadequate the patient should be intubated and ventilated with 100% oxygen. Large bore IVs are established in each antecubital fossa and simultaneously blood drawn for type and cross match and routine lab work. Primary survey evaluating the patient's overall head, chest, abdominal and extremity status is performed. The specific management of the immediately life-threatening chest injury identified in the primary survey will now be discussed.

Cardiothoracic Surgery, Third Edition, edited by Fritz J. Baumgartner. ©2004 Landes Bioscience.

Fig. 19.1. Left tension pneumothorax in a boy with blunt tracheobronchial disruption. Mediastinal shift is clearly evident. A right sided pneumothorax is present as well and is decompressed with a chest tube.

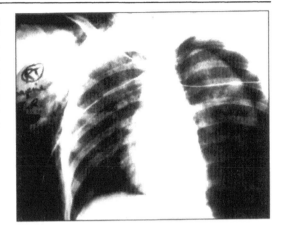

AIRWAY OBSTRUCTION

Management of life-threatening airway obstruction from trauma varies depending on the type of trauma. For example massive oral/facial injuries preventing adequate oral tracheal intubation may require a cricothyroidostomy, whereas injuries to the trachea with suspected tracheal disruption may require fiberoptic bronchoscopic assessment and passage of an endotracheal tube past the tracheal injury using the bronchoscope.

TENSION PNEUMOTHORAX

Tension pneumothorax (Fig. 19.1) may occur either from blunt or penetrating trauma and can be identified by tracheal deviation, unilateral loss of breath sounds, distended neck veins or respiratory distress. If the patient is immediately decompensating from a tension pneumothorax and has no breath sounds on this side of the injury, a large bore needle should be inserted into the second intercostal space in the midclavicular line. If this results in a massive hiss of air, one can conclude that the tension pneumothorax was the cause of the problem. It should be noted that the needle should not be placed too medial, (i.e., parasternal) lest the internal mammary artery be injured.

Should a chest tube be placed for all patients with pneumothorax after blunt or penetrating trauma? The answer is an unequivocal yes. This is because these pneumothoraces frequently enlarge when the patient is stabilized, and if the patient goes to the operating room and undergoes intubation and positive pressure breathing, then the pneumothorax will surely increase and cause compression of the mediastinum with tamponade. The tube is placed laterally and directed posteriorly so as to drain any blood that may accumulate. Although some may advocate that for pneumothorax the tube should be placed anteriorly, generally for trauma it is placed laterally or posteriorly. If there is a tiny amount of effusion or blood present on the initial chest x-ray, there may be justification for not placing a chest tube and watching it expectantly, if there is no evidence of pneumothorax.

OPEN PNEUMOTHORAX

An open sucking pneumothorax occurs with a large open wound. Equilibration occurs between intrathoracic pressure and atmospheric pressure. This results in ineffective ventilation because air does not go into the lung, rather air is sucked into the pleural space from the outside environment. This may be managed by immediate closure of the defect with a large vaseline gauze, large enough to overlap the wounds edges. It is taped firmly in place on three sides, leaving the first side open to result in a one-way valve, such that any air that is trapped inside the pleural space would be extruded but air is not allowed to suck in from the outside environment into the pleural space. A chest tube should be placed and expeditious operative management performed to close the chest wound.

HEMOTHORAX

Massive hemothorax by definition is an immediate loss of 1500 cc or more of blood into the chest cavity and may occur from blunt or penetrating trauma. A cell saver Hemovac type suction apparatus, in which blood can be transfused immediately back into the patient is helpful in this situation. Operation should be performed for such exsanguinating blood loss early in the patient's management. If the patient is hypotensive, immediate operation is needed, sometimes even by thoracotomy in the emergency room if the patient is agonal.

Blood loss exceeding 1000 cc over the next hour or more than 500 cc each hour over the next 2 hours or more than 300 cc each over the next 3 hours generally requires thoracotomy, since a surgically correctable problem is usually identified.

FLAIL CHEST

A flail chest occurs when a portion of the chest wall does not have bony continuity with the remainder of the thoracic cage, i.e., a rib is fractured in two different places and multiple ribs are involved. This leaves the chest wall unstable and causes it to move paradoxically. The hypoxia associated with flail chest is less a result of the actual mechanical paradoxical motion, and clinically the most significant problem is the underlying pulmonary contusion.

Previous historical recommendations have included placing towel clips over the chest to try to stabilize the chest, but generally if one wants to simply stabilize the chest then rolling the patient onto the affected side would be just as effective in preventing the paradoxical motion. Realistically, however, this should rarely, if ever, be done, and the patient should only undergo endotracheal intubation for effective oxygenation and to stabilize the chest.

CARDIAC TAMPONADE

Cardiac tamponade is a great imitator and may fool the physician since it is sometimes difficult to diagnose. Classic findings are distended neck veins, hypertension, plethora from impeded venous return and obvious distress. Tamponade usually results from penetrating injuries although blunt injury may also result in tamponade. A constellation of signs includes plethora, venous pressure elevation, hypotension, tachycardia, muffled heart tones and pulsus paradoxicus. Pulsus paradoxicus is defined as a decrease in the systolic blood pressure by at least 10 mm on inspiration. This is usually not clinically identifiable in the emergency room because of loud emergency room noise, but should be kept in the back of one's mind when evaluating such a patient. The physiologic basis for this phenomenon is that with inspiration, there is increased venous drainage to the right side of the heart which shifts the interventricular septum towards the left side. This results in decreased volume of the left ventricle and decreased cardiac output. Normally, the heart is able to distend within the pericardium, but in the tamponade situation, the heart is compressed from the outside and therefore respiration may effect the blood pressure. This, of course, is true only for those breathing spontaneously, as mechanical ventilation is positive pressure breathing, rather than the normal negative pressure of spontaneous ventilation.

The diagnosis of cardiac tamponade can be very elusive, as patients who initially present with shock may have had massive blood loss, and despite the fact that they also have tamponade, they may not have distended neck veins because of the volume loss. Conversely, cardiac tamponade may be overdiagnosed such as in a young trauma patient who has volume resuscitation resulting in massive over distention of his neck veins. In addition, tension pneumothorax may mimic cardiac tamponade. This is particularly true on the left side where a massive pneumothorax may distort the mediastinum such that there is diminished venous drainage into the heart, thus mimicking the physiologic consequences of cardiac tamponade.

The management of cardiac tamponade is taught by many to include pericardiocentesis. However, it must be emphasized that pericardiocentesis is, at best, a dangerous procedure which may or may not have any benefit toward relieving a tamponade. If tamponade is present in the pericardium, this may have been converted to a blood clot such that it cannot be drawn through the pericardiocentesis needle. Additionally, if there is no tamponade present and the needle is passed into the pericardium and into the heart, one may get a rush of blood back and think that this is decompression of a tamponade when, in fact, it

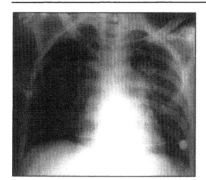

Fig. 19.2. Chest x-ray of a patient with a stab wound to the right ventricle in cardiac tamponade. The cardiac silhouette is not enlarged. The patient underwent successful repair of the injury.

19

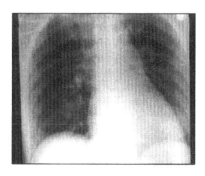

Fig. 19.3. Chest x-ray of a patient in tamponade from blunt avulsion of the inferior vena cava from the right atrium. The patient arrested and required subxiphoid decompression of the tamponade in the ER. Sternotomy and "sucker bypass" to the femoral artery permitted successful repair of the injury. Courtesy Dr. Michael Janusz, University British Columbia.

is actually ventricular blood. There is an additional risk of laceration of the coronary arteries. We feel that pericardiocentesis should rarely, if ever, be done in the trauma situation. For suspected cardiac tamponade with hemodynamic compromise, the patient should undergo either a subxiphoid window (if necessary in the emergency room but preferably in the operating room), or if the patient has fully decompensated, then a left anterolateral thoracotomy should be done with decompression of the tamponade.

The most important aspect of managing a cardiac tamponade is to volume load the patient, and to increase the preload of the heart so as to offset the increased pressure on the outside of the heart. This cannot be overemphasized. Volume must be given to buy enough time to bring the patient to the operating room so that definitive management can be done.

The initial chest x-ray after blunt or penetrating cardiac injury with tamponade may be completely normal and not reveal an enlarged cardiac silhouette (Figs. 19.2 and 19.3). It is frequently assumed by many physicians that injury of the heart from blunt or penetrating trauma with cardiac tamponade will result in dilation of the cardiac shadow. This is completely false since the pericardium is nondistensible and this, in fact, is part of the physiology of cardiac tamponade. In a distressed trauma patient with distended neck veins, tamponade should be suspected despite a completely normal cardiac shadow.

AORTIC RUPTURE

Injuries identified in the secondary survey require a further, indepth physical examination, upright chest film, arterial blood gas and EKG. Traumatic rupture of the aorta most commonly occurs just distal to the ligamentum arteriosum in the proximal descending thoracic aorta. It may range from a small intimal tear of the aorta to a full-blown complete transection with either exsanguinating hemorrhage or preservation of the patient's life by a column of peri-adventitial connective tissue, maintaining the continuity of the aortic lumen. Ninety percent of traumatic ruptures of the aorta are fatal at the time of the accident. Survivors of this event can usually be saved if the aortic rupture is identified expeditiously. Of the 10% who survive the aortic injury after the initial accident, one-half will die each day in the hospital if left untreated. Some of these patients may actually live for years with a traumatic transection of the aorta and not present until much later with a progressive false aneurysm. The radiologic findings of a traumatic transection of the aorta includes blunting of the aortic knob, compression of the left mainstem bronchus, apical capping and widening of the mediastinum. Of these, widening of the mediastinum and blunting of the aortic knob are the two most sensitive indicators (Fig. 19.4a-b). It is important that if these two findings are identified on a supine chest x-ray, that an upright chest x-ray be obtained as well. In most cases, this simple maneuver will result in the appearance of a much narrower mediastinum and a sharp aortic knob. There may be cases of a deceptively sharp aortic knob appearing in aortic rupture (Fig. 19.5a-b). Index of suspicion and mechanism of injury alone may be sufficient indication for an aortogram. Another important radiologic finding is deviation of a nasogastric tube far to the right in patients with traumatic injuries to the aorta because of the large hematoma that develops in the left side of the mediastinum, shifting the mediastinum to right. A clinical finding is diminution of the left upper extremity blood pressure compared to the right upper extremity, resulting from compression of the left subclavian artery from hematoma. The management of these injuries is surgical, using single lung ventilation to collapse the lung, and dissection of the aorta with clamping proximal to the injury between the left subclavian and left common carotid, clamping of the left subclavian and clamping inferiorly. A primary repair is achieved by placing an interposition graft between the fractured segments

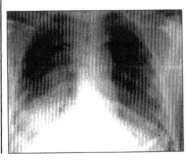

Fig. 19.4a. Upright CXR of a trauma patient who was the unrestrained driver in an motor vehicle accident. The mediastinum is widened and aortic knob blunted.

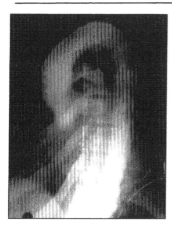

Fig. 19.4b. The aortogram in this patient reveals a complete transection of the aorta. Aortic continuity is preserved only by periaortic connective tissue. The patient underwent uneventful aortic tube graft replacement using left atrial-femoral bypass. Courtesy Dr. Michael Janusz, University British Columbia.

19

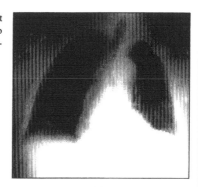

Fig. 19.5a. Upright CXR of a trauma patient sustaining a fall from a tree. The aortic knob is deceptively sharp, although the mediastinum is somewhat widened.

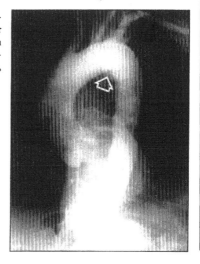

Fig. 19.5b. The aortogram in this patient reveals a pseudoaneurysm from an aortic tear in the region of the ligamentum arteriosum (arrow). Repair was done using left atrial-femoral bypass. Courtesy Dr. Michael Janusz, University British Columbia.

Fig. 19.6a. Aortogram of unrestrained driver in a motor vehicle accident. The aorta is completely transected.

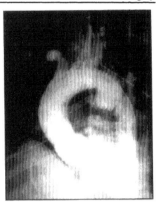

(Fig. 19.6a,b). Spinal protection methods include femoral-femoral bypass or left atrial-femoral arterial bypass. In general, a clamp and go method is feasible if done within 20-25 minutes and usually does not result in paraplegia.

TRACHEOBRONCHIAL INJURIES

Tracheobronchial tree injuries may occur all the way from the larynx down to the subsegmental bronchus. Generally, segmental and subsegmental bronchial injuries do not require treatment but laryngeal, tracheal, mainstem bronchial or lobar injuries do require treatment. If injury to the larynx results in hoarseness, subcutaneous emphysema with crepitus and noisy breathing, the patient should undergo emergency flexible bronchoscopy with an endotracheal tube inserted over the bronchoscope. Thus, when the injury is identified and the bronchoscope is placed past the injury, the endotracheal tube can be positioned under direct vision with the bronchoscope used to stent the airway.

Injuries to the trachea, such as laryngeal injuries, should be managed in a similar way. These injuries are usually from unrestrained passengers who crash their car and smash their neck against the top of the steering wheel resulting in tracheal

Fig. 19.6b. The patient underwent interposition tube graft replacement of the aortic tear. The upper vessel loops are retracting the phrenic and vagus nerves. Repair was done with the "clamp and go" method.

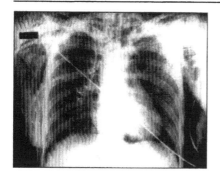

Fig. 19.7. CXR of a young woman sustaining blunt chest trauma in an MVA. There is massive subcutaneous emphysema tracking along the fibers of the pectoralis muscles. A right-sided pneumothorax is decompressed with a chest tube. She was found to have a transection of the right mainstem bronchus and underwent definitive repair.

19

injury. The most common location of the tracheobronchial injuries from blunt trauma occurs within 2-1/2 cm of the carina, i.e., usually a mainstem bronchial injury. Symptoms include respiratory distress, subcutaneous, emphysema and crepitus. It is important to remember that the clinical presentation of tracheobronchial injuries and blunt trauma vary, and the initial diagnostic evaluation is often misleading. Early intervention is important because 50% of deaths from this injury occur within an hour after the injury. Also, if the patient does survive days after the initial injury, the chance of subsequent serious complications—including persistent atelectasis with pneumonia and bronchiectasis, empyema, and stenosis and stricture of the injured segment—is extremely high. One point is that all patients with major tracheobronchial injuries will eventually develop a pneumothorax or subcutaneous emphysema (Figs. 19.7, 19.8). Sometimes this is a delayed finding, and the initial presentation may not include a pneumothorax or subacute emphysema (Fig. 19.9).

Initial management should be directed at securing the airway either by cricothyroidostomy or endotracheal intubation over a flexible bronchoscope. A chest tube is inserted if there is evidence of pneumothorax. If there is a large air

Fig. 19.8a. CXR of a boy in a bicycle versus auto accident 3 days earlier. Initial bilateral pneumothoraces were treated with chest tubes, with persistent air leaks. On the third day, complete collapse of the left lung is demonstrated on the CXR.

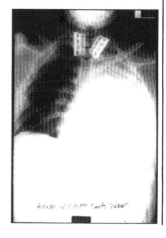

Fig. 19.8b. CT scan. The patient was found to have a major left bronchial injury.

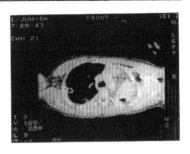

leak with subcutaneous emphysema, a tracheobronchial injury should be suspected. At this point, flexible fiberoptic bronchoscopy should be performed, preferably in the operating room to identify a tracheobronchial injury. If one is identified, definitive surgery should be done at that time. If no tracheobronchial injury is identified on flexible bronchoscopy but there is a massive air leak and a large amount of subcutaneous emphysema, other techniques should be used. These include rigid bronchoscopy and bronchography (Fig. 19.10). When the injury is identified, early operative management is indicated.

The incision depends on the location of the tracheobronchial injury. If the trachea is injured, a cervical incision is usually adequate for injuries in the upper half of the trachea. For injuries in the lower half of the trachea or in the region of

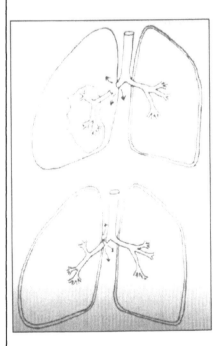

Fig. 19.9a. The most common presentation of bronchial rupture includes a pneumothorax. This is immediately evident if the rupture is intrapleural.

Fig. 19.9b. If the rupture is extrapleural, the air may track within the mediastinum but may not initially present as a pneumothorax.

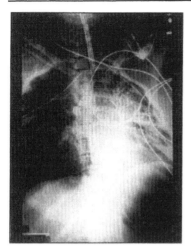

Fig. 19.10. Bronchogram of a patient sustaining a left mainstem bronchial tear. The injury was initially missed on flexible bronchoscopy. Bronchography revealed occlusion of the left lower lobe bronchus. Subsequent bronchoscopy confirmed a spiral tear in the left mainstem bronchus occluding the lower lobe orifice.

the carina or mainstem bronchi, a right posterolateral thoracotomy is performed through the fourth intercostal space which gives good access to the region of the carina and lower trachea. It is important to remember that access to the left mainstem bronchus can also be achieved from the right side. If it is known that the only bronchial injury is on the left side, then a left posterolateral thoracotomy is justified.

ESOPHAGEAL TRAUMA

Esophageal trauma may occur concomitant with tracheobronchial injury. This typically manifests several days after the tracheobronchial injury from necrosis of the anterior portion of the esophageal wall resulting in a traumatic tracheoesophageal fistula from blunt trauma. Esophageal trauma, however, is most commonly caused by penetrating trauma. Usually this is the result of a transmediastinal gunshot wound and may be difficult to diagnose. The overall management of transmediastinal gunshot wounds will be reviewed; however, after major vascular injuries have been ruled out, the patient should undergo rigid esophagoscopy followed by a Gastrografin swallow. If one is still suspicious after a negative Gastrografin swallow, then a barium swallow should be done. If an esophageal injury is missed, this can lead to lethal consequences from empyema. Management of an esophageal injury is early thoracotomy and primary repair of the injury with placement of a pleural flap over the injury. For the lower portion of the esophagus, exposure is through the left chest. For the mid-esophagus (i.e., upper thoracic esophagus) a right thoracotomy is done.

PULMONARY CONTUSION

The main problem that occurs with pulmonary contusion is inadequate ventilation from injured parenchyma. Severe pulmonary contusion should be managed with endotracheal intubation and ventilation. This may not always be necessary and the patient can sometimes be managed with pulmonary toilet including percussion and postural drainage, bronchodilator therapy and deep breathing exercises.

MYOCARDIAL CONTUSION

This has been documented by careful studies at Harbor-UCLA to have greatly overrated clinical significance. While it is true that blunt cardiac injury can lead to rupture of the heart which is usually immediately evident, nonrupture of the heart from blunt trauma usually will not result in major life-threatening hemodynamic sequelae. It is true, however, that on occasion there will be a patient with arrhythmias resulting from blunt injury to the heart that must be managed with anti-arrhythmic therapy. In general, however, a patient who has blunt chest trauma who has no evidence of EKG changes on his emergency room EKG has an extremely low risk of developing future arrhythmias related to his myocardial contusion. Right ventricular dysfunction may occur since the right ventricle is the most anterior portion of the heart and more susceptible to compression from a direct sternal blow. MUGA scan or cardiac ECHO are usually not indicated, however, unless there are specific indications to do so.

TRAUMATIC DIAPHRAGM RUPTURE

Traumatic diaphragmatic rupture is more common on the left side in a similar way that the congenital diaphragmatic Bochdalek's hernia occurs on the left side. It is frequently diagnosed when the nasogastric tube in a trauma patient enters the stomach which is found to be in the left chest. Injuries may be missed initially when the x-ray is misread as an elevated left hemidiaphragm or dilated stomach bubble. Simply passing a nasogastric tube, however, will confirm the diagnosis, or if there is still question an oral contrast agent may be given. Exploratory laparotomy and repair of the diaphragmatic remnant will cure the problem.

BONY CHEST FRACTURES

Bony fractures of the chest wall in and of themselves do not require treatment. They may be related to other injuries including hemothorax, pulmonary contusion, pneumothorax or other cardiac or vascular injuries. In particular, there is an association between scapular fractures and first rib fractures with the presence of injuries to the aorta. As a general rule, younger patients with flexible chest walls are less likely to have rib fractures.

19

INJURIES TO THE CARDIAC VALVES AND CORONARY ARTERIES

Blunt and penetrating trauma can result in injuries to the coronary arteries or cardiac valves. The left anterior descending artery (LAD), because of its more anterior location, is the most susceptible to injury. Injuries to the distal third of the LAD can generally be managed with ligation of the vessel. More proximal injuries generally require cardiopulmonary bypass and reverse saphenous vein grafting to the distal LAD (Fig. 19.11a-b). Penetrating injuries to the heart may result in valvular injuries which may be overlooked in the haste of repairing an external cardiac injury. Transesophageal echocardiography is invaluable in assessing valvular function in the presence of penetrating cardiac trauma. Blunt rupture of cardiac valves has also been reported, with successful repair performed.

TRANSMEDIASTINAL GUNSHOT WOUNDS

This is a complex problem because of the variety of injuries that may occur in the chest from vascular or cardiac injury, aeroesophageal injury or spinal injury. Priority lies with life-threatening cardiac and vascular injuries. After the initial resuscitation, if the patient remains stable, he should be transported to the angiography suite for arteriogram of the aortic arch vessels and ascending aorta (Table 19.1). If this is negative and the patient is still stable, he should go to the operating room for endotracheal intubation followed by flexible bronchoscopy and rigid esophagoscopy. If these are negative, the patient is taken back to the radiology suite for a Gastrografin swallow to rule out esophageal injury, and if this is negative then an esophageal swallow with barium if there is still high suspicion for esophageal injury. The reason that a radiologic swallow study should generally be done after a negative rigid esophageal endoscopy is that the radiologic study is actually more sensitive than the endoscopic study. Of course if at the time of surgery esophageal endoscopy reveals an injury then it should be managed operatively either by a right or left thoracotomy, or if in the cervical region a cervical approach. Tracheobronchial injuries are likewise managed depending on the location of the injury.

COMBINED CHEST, ABDOMINAL AND HEAD TRAUMA

Trauma involving the chest, abdomen and brain requires aggressive and skillful diagnosis and management. General priorities can be established. Abdominal trauma should take precedence over all other trauma. A surgically correctable problem in the abdomen is a much more common cause of shock than a surgically

Fig. 19.11a. Intraoperative photograph of a young man sustaining a gunshot wound to the proximal LAD (arrow). The left ventricle rapidly became hypo- and then dyskinetic, and cardiopulmonary bypass was established.

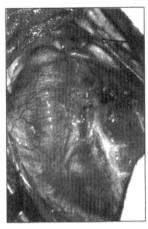

Fig. 19.11b. A reverse saphenous vein graft to the distal LAD was performed (arrow). A pledgetted stitch is seen at the LAD injury (large arrow).

Table 19.1. Thoracic trauma; transmediastinal GSW

Management of stable patient

- First, arteriography

- Next, to OR for flexible broncoscopy and rigid or flexible esophagoscopy

- Next, Gastrografin esophageal swallow

correctable problem in the chest or brain. Another caveat is that the patient must be accompanied by the trauma surgeon at all times, even if the blood pressure is normal because a stable blood pressure does not necessarily imply a stable patient. If a surgical lesion is found on peritoneal lavage or CT scan of the abdomen, whether the patient is stable or unstable, exploratory laparotomy should be done first and has priority over any possible chest injury. Another important point is that a CT scan of a traumatic thoracic aortic injury may be misleading and the gold standard is still aortography.

In the operating room, the left chest should be widely prepped and draped when doing an exploratory laparotomy in the event that the patient suddenly decompensates hemodynamically in a way that cannot be explained by the intra-abdominal injury. At that point, a left thoracotomy can be done to inspect the chest and aorta. If, however, a massive hemothorax presents with instability, the chest injury takes precedence. The general philosophy is that a correctable life-threatening abdominal injury is more common than a correctable life-threatening chest injury. The above considerations apply to combined chest and abdominal trauma.

For cranial injury, the same general philosophy holds, but it must be kept in mind that if the patient decompensates in the operating room during laparotomy, the decision whether to and where to perform a craniotomy is not as simple as a left thoracotomy. Thus the diagnostic priorities must reflect this, for even though abdominal trauma takes priority over brain trauma, it still is extremely helpful to have the CT scan of the brain available if the patient decompensates.

Complex trauma involves combined head, chest and abdominal trauma. Abdominal trauma takes precedence over chest trauma with the exception of a massive hemothorax with hypotension in which the patient has persistent exsanguinating hemorrhage from the chest. The following comments refer to a situation in which there is not an obvious massive thoracic hemorrhage. In an unstable patient with all three types of injuries, if the patient has a tense distended abdomen, he should go to the operating room for a laparotomy for correction of the abdominal hemorrhage. This should then be followed by head CT scan, followed by a craniotomy if necessary. After the above are accomplished, only then should an arch angiogram be obtained if there is reason to do so, i.e., blunt chest trauma with a widened mediastinum or blurring of the aortic knob. On the other hand, if the patient has a flat abdomen without an obvious bleed into the abdomen and is unstable, then a peritoneal lavage should be performed in the emergency room because sometimes exsanguinating hemorrhage may occur in the abdomen without a tense, distended abdomen. If the patient is neurologically compromised from head trauma, the abdominal exam will be unreliable. If the peritoneal lavage is positive, the patient should go to the operating room for laparotomy. This again is followed by head CT scan, plus or minus a craniotomy, followed finally by an arch angiogram if indicated. If the peritoneal lavage is negative, the patient should undergo a head CT scan plus or minus craniotomy, and then an arch angiogram if indicated based on the type of injury and chest x-ray.

Another category is a stable patient without abdominal findings who has combined abdominal head and chest trauma and who is neurologically compromised. After the primary and secondary survey, CT scan of the head is obtained. If there is evidence of a neurosurgically correctable injury, a craniotomy is performed. In the operating room a lavage is done to rule out an abdominal injury, and if this is positive, a laparotomy is performed. After this, the patient undergoes an arch

Table 19.2. Combined head, chest, abdominal trauma

COMBINED HEAD, CHEST, ABDOMINAL TRAUMA

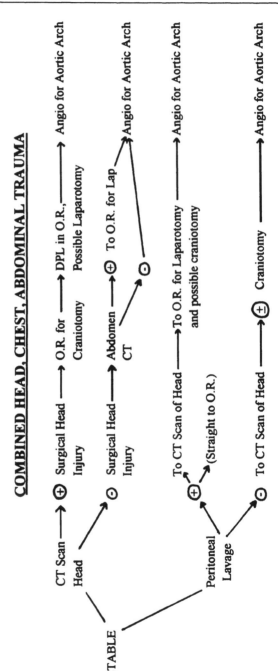

19

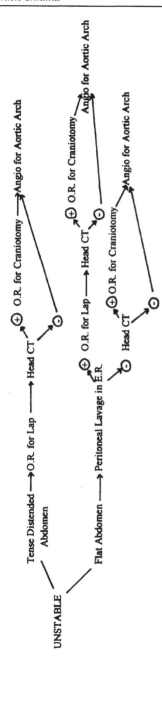

GENERAL PRIORITIES: ABDOMEN, BRAIN, CHEST

angiogram if indicated based on the type of injury and the chest x-ray. If the original CT scan of the head does not reveal a neurosurgically correctable injury, an abdominal CT scan is then done while the patient is in the CAT scanner. If the CT scan is positive, the patient is taken to the operating room for laparotomy and this is then followed by an arch angiogram if indicated based on the mechanism of injury and chest x-ray. If the abdominal CT scan is negative, then again an arch angiogram is done based on the above indications.

Table 19.2 depicts the management sequence for combined abdominal, head and chest trauma in stable and unstable patients. Again it should be emphasized that this refers only to patients who do not have obvious exsanguinating hemorrhage from the chest and do not require immediate thoracotomy. This method of managing these complex patients cannot be taken as a hard and fast rule, since individualization and management for these particular patients is necessary at all times.

SUGGESTED READING

1. Wiot JF. The radiologic manifestations of blunt chest trauma. JAMA 1975; 231:500.
2. Defore WW, Mattox KL, Hansen HA et al. Surgical management of penetrating injuries of the esophagus. Am J Surg 1977; 134:734.
3. Mattox KL, Pickard LR, Allen MK et al. Suspected thoracic aortic transection. J Am Coll Emerg Physicians 1978; 7:12.
4. Blair E, Topuzulu C, Deane RS. Major chest trauma. Current Problems in Surgery 1969; 2-69.
5. Jones KW. Thoracic Trauma. Surg Clinics N. American 1980; 60:957.
6. Richardson JD, Adams L, Flint LM. Selective Management of Flail chest and Pulmonary Contusion. Ann Surg 1982; 196:481-487.
7. Baumgartner F, Sheppard B, de Virgilio C et al. Tracheal and Main Bronchial Disruptions after Blunt Chest Trauma: Presentation and Management. Ann Thorac Surg 1990; 50:569-574.

Congenital Causes
of Respiratory Distress

20

Fritz J. Baumgartner

This is an important topic since in general these are well-defined, surgically correctable lesions. The surgically correctable causes of respiratory distress in the newborn include bronchogenic cyst, Bochdalek's hernia, sequestration (usually intralobar), congenital lobar emphysema and cystic adenomatoid malformation. Resuscitation of the infant should be done with oxygen and intubation if necessary. Chest x-rays are important since they can be diagnostic in certain instances.

Bronchogenic cysts are lined with ciliated respiratory cells and can be found in the hilum, lung or mediastinum. Surgery should be performed for all bronchogenic cysts since they may compress the airway or become infected. Bronchogenic cysts may be found in adulthood as well (Fig. 20.1).

Bochdalek's hernia involves a defect in the posterolateral portion of the diaphragm, usually on the left, and the x-ray can be diagnostic since it shows intestine in the newborn's chest cavity compressing the lung with a shift of the mediastinum towards the right. Management includes intubation and resuscitation and immediate surgical intervention. The herniated bowel is reduced with repair of the diaphragm. The lungs may be hypoplastic from in utero compression of the lungs by bowel. A period of ECMO may be worthwhile in these situations if there is poor oxygenation despite intubation and oxygen therapy.

Sequestration consists either of the more common intralobar variety or the less common extralobar variety. The intralobar variety usually is in the left lower lobe, receives a systemic arterial supply usually directly off the aorta, drains by the pulmonary vein, and usually has a patent bronchus (Fig. 20.2). Lobectomy is the treatment of choice for intralobar sequestration. Extralobar sequestration likewise is primarily in the left lower lobe. There is usually no patent bronchus present. Venous drainage is via the azygous venous system. There are frequently other congenital anomalies present. No lobectomy is necessary since the incidence of pneumonia and pulmonary destruction is much less common because there is no patent bronchus present.

To differentiate intralobar and extralobar sequestration, CT scan may be necessary as may an arteriogram. Congenital lobar emphysema involves an isolated hyperinflation of the lobe without extrinsic obstruction. Symptoms of respiratory distress exist usually in the first week but may extend over several months. It is more common in males than females. The symptoms are tachypnea, dyspnea, cyanosis and cough. Usually the left upper lobe is involved. In 25%, there is carti-

Fig. 20.1a. Gross specimen of a large bronchogenic cyst of an adult male. A large amount of infected mucus and clot was retrieved from the cyst. Courtesy Drs. Bill Nelems and Ken Evans, University of British Columbia.

Fig. 20.1b. The cut specimen shows the cyst in relationship to the lung parenchyma. Courtesy Drs. Bill Nelems and Ken Evans, University of British Columbia.

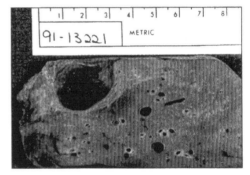

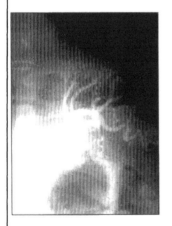

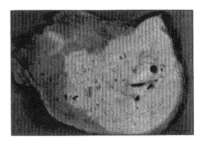

Fig. 20.2a. Angiogram of the systemic arterial supply of an intralobar left lower lobe sequestration. Courtesy Drs. Bill Nelems and Ken Evans, University of British Columbia.

Fig. 20.2b. A left lower lobectomy was performed revealing the large sequestration. Courtesy Drs. Bill Nelems and Ken Evans, University of British Columbia.

laginous dysplasia of the lobe. In older children, bronchoscopy should be done to rule out foreign body. After intubation, the chest should be opened as soon as possible since positive pressure breathing will further hyperinflate the lung and may cause hemodynamic compromise. Cystic adenomatoid malformation involves overgrowth of the terminal bronchioles and lack of mature alveoli resulting in multiple cysts with both solid and cystic components. Respiratory distress may exist at birth. It may be confused radiographically with a Bochdalek's hernia.

SUGGESTED READING

1. Cullen ML, Klein MD, Philippart AI. Congenital diaphragmatic hernia. Surg Clin North Am 1985; 65:1115-1138.
2. Eigen H, Lemen RJ, Waring WW. Congenital Labor emphysema: Long-term evaluation of surgically and conservatively treated children. Am Rev Resp Dis 1976; 113:823-827.
3. Flye MW, Conley M, Silver D. Spectrum of pulmonary sequestration. Ann Thorac Surg 1976; 22:478-482.
4. Haller JA Jr, Golladay ES, Pickard LR et al. Surgical management of lung bud anomalies; lobar emphysema, bronchogenic cyst, cystic adenomatoid malformation, and intralobar pulmonary sequestration. Ann Thorac Surg 1979; 28:33-43.

Chest Wall Congenital Deformities

Fritz J. Baumgartner

Congenital deformities of the chest wall include pectus excavatum and pectus carinatum. The most common congenital deformity is the excavatum variety in which the sternum is displaced posteriorly. Asymmetry is common and almost always involves deeper depression of the right costal cartilage with rotation of the sternum to the right. Most patients are asymptomatic; however, some patients have symptoms including chest pain, dyspnea, poor feeding, bronchospasm and arrhythmias. Most of these symptoms are relieved by operative correction. It is controversial whether or not there is actual impairment in exercise cardiac function. The data at present show no impairment in exercise cardiac function before or after pectus repair. It has been shown, however, that left ventricular end diastolic volume and stroke volume increase at rest after surgery and there is a marked increase in resting right ventricular end diastolic volume. This suggests that some degree of cardiac compression was relieved by repair. Most surgeons recommend operative correction before the age of five, but not before age one and a half. A submammary incision is performed. All deformed costal cartilages are excised subperichondrially. The xiphoid sternal joint is disarticulated and a transverse sternal osteotomy is performed above the point of depression. The sternum is brought forward and a steel bar is placed retrosternally and anchored at the ribs.

Although it is difficult to evaluate, in general pulmonary function tests and stroke volume in patients with pectus excavatum are decreased compared to controls and are improved by surgery, although the clinical relevance is unknown at this time.

Pectus carinatum is much less common than excavatum and is an out-pouching deformity of the chest of several varieties. The two types are the chicken breast with a deep depression of the costal cartilage on either side of the sternum. Asymmetry is common usually on the right. Another deformity is the Pouter-pigeon variety which involves a double angle in the sternum with a depression created in the lower sternum resembling a pectus excavatum. Again a submammary incision is performed and a subperichondrial resection of deformed cartilage is made. The excess length of perichondrial bed is plicated and the sternum transsected transversally with an osteotomy if necessary.

SUGGESTED READING

1. Ravitch MM. Congenital deformities of the chest wall and their operative cor-
    rection. Philadelphia: WB Saunders, 1977.
2. Robicsek F, Dougherty HK, Mullen DC et al. Technical considerations in the
    management of pectus excavatum and carinatum. Ann Thorac Surg 1974;
    18:549-564.

21

Thoracic Outlet Syndrome, Hyperhidrosis and Causalgia

GENERAL CHARACTERISTICS OF THORACIC OUTLET SYNDROME

The thoracic outlet syndrome is a constellation of neurovascular symptoms resulting from compression at the thoracic outlet. Ninety-five percent of thoracic outlet syndromes result from neurologic symptoms usually from compression of the lower roots of the brachial plexus. Only 5% of the thoracic outlet syndrome is vascular in etiology. The brachial plexus consists of input from C5, C6, C7, C8 and T1. The most common cause of the thoracic outlet syndrome results from compression of the lower roots of the brachial plexus, i.e., the ulnar distribution contributed to from C8 and T1.

The thoracic outlet is that region bordered by the clavicle and first rib (Fig. 22.1). Posteriorly, it is bounded by the scalene medius and the long thoracic nerve. The anterior scalene inserting on the scalene tubercle of first rib divides the thoracic outlet into an anterior compartment containing the subclavian vein, and the posterior compartment containing the subclavian artery and the brachial plexus. The most common cause of the thoracic outlet syndrome is a cervical rib. This occurs in about 95% of patients with the syndrome. Other bony abnormalities may exist and contribute to the thoracic outlet syndrome. This includes rudimentary first ribs or fusion of the first and second ribs, clavicular deformities or first rib fractures. First rib anomalies are often related to vascular problems related to thoracic outlet syndrome (Fig. 22.2a-c).

There are several tests used to mimic the findings of the thoracic outlet syndrome. Most of these are neither sensitive nor specific and generally are of questionable value but will be included here for completeness. Perhaps the most useful of all these tests is the Roos test.

PROVOCATIVE MANEUVERS

The patient, sitting straight, takes a deep breath and holds it. This maneuver tightens the anterior middle scalene muscles resulting in possible decrease in the radial pulse. Unfortunately, it is a very frequent finding in normal individuals.

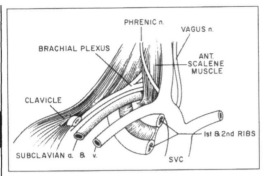

Fig. 22.1. Anatomic relationships at the thoracic outlet.

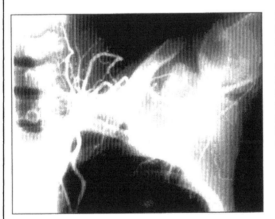

Fig. 22.2a. Left subclavian arteriogram in a patient with thoracic outlet syndrome with the arm abducted and externally rotated demonstrating complete occlusion of the subclavian artery with reconstitution of the axillary artery by extensive collaterals.

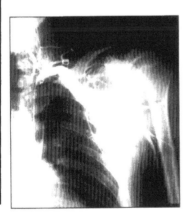

Fig. 22.2b. Left subclavian arteriogram in another patient with the arm adducted demonstrating significant post-stenotic dilation of the subclavian artery.

Fig. 22.2c. When the arm is abducted and externally rotated, an obstruction to the subclavian artery is seen. Further abduction of the arm resulted in complete occlusion at this level. Both these patients had a rudimentary first rib with an abnormal exostosis between the first and second ribs. This was the cause of the compression to the subclavian artery. Transaxillary resection of the first and second ribs was curative in both cases.

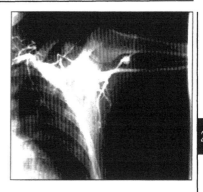

22

COSTOCLAVICULAR TEST (MILITARY POSITION)

In the costoclavicular test (military position) the shoulders are drawn downward and backward, thus narrowing the costoclavicular space by bringing the clavicle close to the first rib. Changes in the radial pulse with symptoms indicates compression.

ROOS TEST

The Roos test is the most useful test. The arms are raised to the horizontal position and the elbows flexed to 90°. The hands are rapidly opened and closed. If numbness or pain is elicited in the hands and forearms, this is a sign of compression and thoracic outlet syndrome.

The history and physical examination is critical in the evaluation of thoracic outlet syndrome. Chest x-rays, concentrating on the possibility of cervical ribs or first rib abnormalities are essential. Electromyogram and ulnar nerve conduction velocity is controversial; however, they should be done. The test may be normal in effective individuals; nonetheless an abnormal test may be a more compelling indication that thoracic outlet syndrome does exist. If there are symptoms of vascular compromise, an angiogram should be obtained. A venogram may be required to evaluate the possibility of subclavian vein thrombosis. If the thoracic outlet syndrome involves arterial vascular compromise, surgery is mandatory, since physiotherapy will do little to relieve the symptoms of the vascular problem. However, if there is neurologic compromise only, a trial of physiotherapy is required. This involves various physical therapy exercises, including shoulder strengthening and arm exercises, as well as strengthening of the upper trapezius and proper posture. Sagging of the shoulder girdle, common among the middle-aged, is a major factor in the syndrome.

In the past, one of the primary surgical modalities for treating thoracic outlet syndrome was transection of the anterior scalene muscles or scalenotomy. This, however, was found to be of limited benefit and in general should only be used for "upper nerve" symptomatology, i.e., that involving the C5, 6, 7 nerve roots. The lower nerve roots, C8-T1, really are helped little by a scalenotomy, and the more effective treatment is resection of the first rib. It is very important that this resection be complete since partial first rib resections are known to cause recurrent symptoms of thoracic outlet syndrome. The surgical approach to the first rib is usually transaxillary.

TRANSAXILLARY FIRST RIB RESECTION

The technique of transaxillary first rib resection is as follows: The surgeon stands anterior to the patient. The patient is placed in the lateral decubitus position with the arm held upward on a pulley. An axillary incision is made at the base of the hairline down to the chest wall which is explored with the finger going all the way up to the first rib, avoiding the intercostal brachial nerve and the long thoracic nerve. The long thoracic nerve is avoided by not cutting the posterior portion of the serratus anterior. A periosteal elevator is then used to go underneath the first rib and clean off its inferior surface. The anterior scalene muscle, inserting on the first rib, is transected followed by the scalene medius, so the subclavian artery, vein and brachial plexus are avoided.

The entire first rib is cleaned anteriorly and posteriorly to the transverse process of T1. The first rib is then transected and removed. It is important to remember that the thoracic outlet syndrome may be a very difficult diagnosis to make and the differential diagnosis includes problems of the cervical spine with radicular pain, as well as brachial plexus problems such as superior sulcus tumors invading the lower roots of the brachial plexus. Also carpal tunnel syndrome may mimic thoracic outlet syndrome, as may other peripheral neuropathies.

If a cervical rib is present and is causing the thoracic outlet syndrome, then resection of the cervical rib can be accomplished along with the first rib resection via the transaxillary approach. A scalenotomy is no longer performed; rather a scalenotomy should be done along with neurolysis of C5, C6 and C7 is upper brachial plexus symptomatology exists.

CAUSALGIA

Another pain syndrome which will be discussed is post-traumatic vasomotor dystrophy (causalgia), which is not thoracic outlet syndrome but may still be managed by the thoracic surgeon under appropriate circumstances. This is a syndrome of intense burning pain resulting from trauma, in which the patient experiences severe pain long after the episode of the trauma. This may be elicited by such minor things such as cool air blowing on the extremity. A thoracic sympathectomy may be curative at times for these individuals.

HYPERHYDROSIS

Hyperhidrosis is a disorder resulting in debilitating, excess sweating on one, or more commonly both sides of the body due to overactivity of the upper thoracic sympathetic chain. It usually involves the upper extremities, frequently coexistent with hyperhidrosis of the feet, and frequently has a hereditary component. In severe cases conservative measures including topical aluminum chloride compounds (Drysol), oral anticholinergics, iontophoresis (electrical current transmission via tap water) and Botulinum toxin injections are wholly unsatisfactory. The single most effective treatment is thoracic sympathectomy which is curative for palmar hyperhidrosis and frequently markedly improves hyperhidrosis of the feet as well.

The approach for thoracic sympathectomy for either hyperhidrosis or causalgia can be done open under direct vision through the transaxillary approach.

A double lumen endotracheal tube is employed, as is the lateral decubitus position. Transaxillary incision is performed sparing the intercostal brachial nerve and long thoracic nerve, and the third intercostal space is entered. The posterior mediastinal pleura is incised and the sympathetic chain is exposed from the level of T3 to T1. T1 is the lower portion of the stellate ganglia at the top of the first rib. The rami communicantes are individually clamped with clips and transected. It should be noted that only the lower portion of the stellate ganglia, i.e., T1, should be taken because removal of the C8 portion of the stellate ganglia will result in Horner's syndrome, i.e., ptosis, meiosis and anhidrosis of that side of the face.

Alternatively, the entire ganglia and rami at the level of the second and third sympathetic ganglia are cauterized-transected. Patients should be informed of the possibility of compensatory hyperhidrosis affecting areas other than the upper extremities. Video-assisted thoracoscopic techniques may also be utilized for thoracic sympathectomy. Compared with the open procedure, thoracoscopic sympathectomy is faster, less painful, and achieves better visualization of the thoracic sympathetic chain and can commonly be done on an outpatient basis. It is the procedure of choice for surgical sympathectomy. The procedure is described in Chapter 25.

SUGGESTED READING

1. Roos DB. Congenital anomolies associated with thoracic outlet syndrome-anatomy, symptoms, diagnosis and treatment. Am J Surg 1976; 132:771-778.
2. Urschel HD Jr, Paulson DI, McNamara JJ. Thoracic outlet syndrome. Ann Thorac Surg 1968; 6:1-10.
3. Pollak EW. Surgical anatomy of the thoracic outlet syndrome. Surg Gynecol Obstet 1980; 150:97.
4. Baumgartner F, Nelson RJ, Robertson JM. The rudimentary first rib: a cause of thoracic outlet syndrome with arterial compromise. Arch Surg 1989; 124:1090-1092.

Pulmonary Mechanics

Fritz J. Baumgartner

The elastic properties of the lung and chest wall and air flow resistance are related to breathing mechanics. Figure 23.1 shows the lung volumes used in respiratory physiology. Vital capacity is the maximum volume that can be expired after a maximal inspiration. Total lung capacity is the volume of the lungs after a maximal inspiration. Residual volume is that volume remaining in the lungs after a maximal expiration. This is about 25% of the total lung capacity. Functional residual volume is the volume in the lungs at the end of a normal expiration. Tidal volume is the volume of a spontaneous breath. These are the major volumes; the inspiratory capacity, expiratory reserve volume, and inspiratory reserve volume are less commonly used terms.

23

The pressure volume curve defines the elastic properties of the lung. Figure 23.2 shows such a curve. It shows that with fibrosis, more intrapleural pressure is required to establish lung volume whereas with emphysema much less intrapleural pressure is required to inflate the lung. The elastic properties of the lung reside in the alveolar walls, which contain a thin film of surfactant, resulting in surface tension accounting for lung elasticity. The surface tension increases as the size of the alveolus decreases, and alveolar collapse would result if it were not for this surface tension. Thus, surfactant is extremely important in preventing atelectasis.

Certain factors restrict movement of the chest wall and reduce compliance. This includes diaphragmatic compression caused by abdominal distention such as with obesity, ascites, pregnancy or intra-abdominal tumors. These conditions reduce vital capacity, total lung capacity and functional residual capacity. The loss of elastic properties of the lung itself, such as with emphysema or fibrosis, also can effect the elastic properties of the lung as shown in the pressure volume curve.

The FEV-1 is the forced expiratory volume in one second. This is the commonest flow measurement. The FEV-1 is often reported as a percentage of the vital capacity (FEV-1/VC) as well as the natural volume. It is important to note the value both ways. If the vital capacity is significantly reduced, the ratio of FEV-1/VC may be satisfactory even though the actual volume exhaled is abnormal. The generally accepted minimal postoperative FEV-1 value is 0.8 liters. It can be predicted that postoperative FEV-1 of less than this will result in a pulmonary cripple, although this statement needs to be modified under different circumstances depending on the individual. For example, an elderly woman may require less of an FEV-1 than a young male athlete.

Another commonly used test is the maximal voluntary ventilation (MVV) which is effort dependent and performed by having the patient inhale as deeply and rapidly as possible for about 10 seconds. The MVV usually correlates well

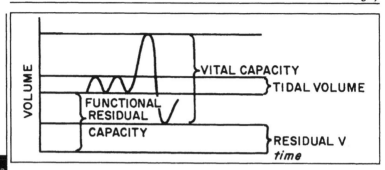

Fig. 23.1. Lung volumes.

Fig. 23.2. Lung pressure-volume curve.

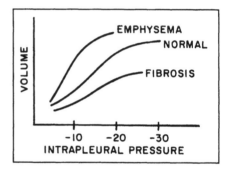

with FEV-1 (FEV-1 x 34 will generally give the MVV). If the actual MVV does not correlate with the calculated value, it suggests poor effort or a tired patient.

Figure 23.3 shows another important test of pulmonary function. This is the flow volume loop. The pressure volume and flow relationship of lung airway obstruction can be determined by measuring maximal expiratory flow since maximal inspiratory flow is relatively independent of effort and primarily dependent on recoil pressure in the lung, which in turn is dependent on lung volume. The measured flow can be related to the lung volume at which it is measured. The classic example is the flow volume loop that results from an obstruction, for example a tracheal tumor or other intraluminal airway obstruction.

Why is pulmonary function worse postoperatively resulting in atelectasis? There are several reasons: (1) pain causing hypoventilation can result in a poor respiratory effort and can lead to hypoventilation; (2) decreased lung volumes including decreased TLC, decreased FRC and decreased end respiratory volume; (3) retained secretions; (4) increased closing volumes.

A closing volume is that volume required to keep the alveoli open and nonatelectatic. Postoperatively, the closing volume may exceed the expiratory reserve volume leading to atelectasis.

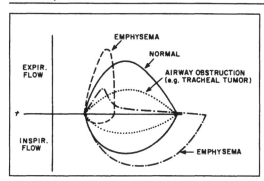

Fig. 23.3. Lung flow volume loop.

SUGGESTED READING

1. Cugell DW, Fish JE, Glossroth J. Mechanics of Breathing. In: Shields TW, ed. General Thoracic Surgery. 3rd ed. Philadelphia: Lea and Febinger, 1989:110-118.
2. West JB. Respiratory Physiology: The Essentials. 2nd ed. Baltimore: Williams and Wilkins, 1979.
3. Rahn H et al. The pressure-volume diagram of the thorax and lung. Am J Physiol 1946; 146:161.

Surgery of the Trachea

ANATOMY

The trachea is a conduit for ventilation. It would seem to be an ideal structure for reconstruction; however there are several unique anatomic features that make it extremely difficult to surgically reconstruct in certain instances. It is a rigid structure and is relatively short and close anatomically to major cardiovascular structures. The adult trachea averages 12 cm in length from the level of the cricoid to the carina. There are approximately twenty cartilagenous rings, about two rings per centimeter. The trachea, when viewed laterally, actually does not proceed in a direct superior-inferior course, but rather courses inferiorly and posteriorly and angles from a nearly subcutaneous position anteriorly going back to rest against the esophagus quite a distance posteriorly. The thyroid isthmus passes over the trachea in the region of the second ring.

The upper portion of the trachea is supplied by the inferior thyroid artery. The lower portion of the trachea is supplied by bronchial arteries. The arteries approach the trachea laterally and devascularization of the trachea is most pronounced when the dissection is performed laterally rather than anteriorly or posteriorly. Lying in the groove between the trachea and the esophagus are the recurrent nerves, coursing around the ligamentum arteriosum on the left and around the subclavian artery on the right. A nonrecurrent nerve occasionally is present on the right side. The nerves enter the larynx between the cricoid and thyroid cartilage. The trachea consists of an anterior cartilagenous portion and a posterior membranous portion. The trachea is lined with respiratory mucosa which is columnar and ciliated. In chronic smokers, squamous metaplasia frequently occurs.

PATHOLOGY

Patients with primary tumors of the trachea may present with upper airway obstruction, hemoptysis, chronic cough or recurrent pneumonia. Stridor may appear on forced respiration and may progress to breathing at rest as well. The most common tumor of the trachea is the squamous cell carcinoma. The second most common tumor is the adenoid cystic carcinoma (cylindroma). About a third of patients with squamous cell carcinoma of the trachea have extensive mediastinal involvement when first seen. Adenoid cystic carcinomas on the other hand usually have not invaded mediastinal structures when first seen. However, these

Cardiothoracic Surgery, Third Edition, edited by Fritz J. Baumgartner. ©2004 Landes Bioscience.

tumors extend greater distances in the trachea wall than is grossly evident. This extension is by submucosal and perineural invasion which is a histologic characteristic of this type of tumor. The trachea is sometimes involved with neoplasms that are metastatic from other sources (e.g., larynx, esophagus, lung or thyroid gland). This is usually by direct extension rather than hematogenous spread. Inflammatory diseases may affect the trachea as well. This may include endotracheal tuberculosis. Strictures were reported following diphtheria as well, but are not seen presently. Sclerosing mediastinitis (e.g., from histoplasmosis) may produce tracheal stenosis as well. Post-intubation damage to the trachea may also occur. High pressure, low volume, cuffed endotracheal tubes were the primary cause of post-intubation damage, however with the new high volumes, low pressure endotracheal cuffs, this is a much less frequent finding. Besides the location of the endotracheal cuff, another possible source of stricture may be the stoma where the tracheostomy tube enters into the trachea. Any patient who has symptoms of airway obstruction who has been previously intubated should be considered to have a possible obstruction secondary to intubation injury.

SURGICAL TREATMENT

With malignant tumors, the goal of surgery is relief of obstruction as well as possible cure. Radiation alone will usually lead to recurrence within several years with squamous cell carcinoma as well as adenoid cystic carcinoma. What if there are localized pulmonary metastases with tracheal squamous cell carcinoma versus adenoid cystic carcinoma? Because of the long clinical course of adenoid cystic carcinoma and less aggresive nature compared to the squamous variety, resecting an obstructing tracheal lesion for adenoid cystic may be justified with limited pulmonary metastases. It is hard to justify this with squamous cell carcinoma, however. It is important in patient selection to avoid the need for ventilatory support postoperatively. This is because in a shortened post-resection trachea, the endotracheal tube cuff may rest against the anastomosis increasing the chance of dehiscence.

With regard to tracheal reconstruction, it should be noted that anatomic mobilization is crucial to achieving an adequate anastomosis that will not dehisce. Mobilization is achieved in one of several ways. The first is simple dissection around the trachea. This should not be extensive and should not extend more than 2 cm beyond the proposed margin of the resection. In particular, it is important to avoid dissecting in the lateral plane of the trachea so as not to devascularize it. Another extremely important maneuver is to flex the neck which will deliver almost the entire cervical trachea into the mediastinum. Thus all patients who undergo tracheal resection should have their neck flexed and their chin sutured to the anterior chest wall. Another way to further immobilize the trachea is to free the hilus of the right lung and to divide the inferior pulmonary ligament. The hilar mobilization is accomplished by freeing the pulmonary vessels from their pulmonary attachments. These maneuvers are most useful during surgery of the lower half of the trachea where generally a right thoracotomy is performed. It has been reported that these maneuvers as well as cervical flexion will allow up to 5 to 6 cm of trachea to be removed by the transthoracic approach. Another maneuver that may improve mobilization is the laryngeal release. In this approach, which is shown in Figure 24.1a-c, the attachments holding the larynx to the myelohyoid are transected. This is accomplished by transecting the muscle attachments to the superior surface of the hyoid, then transecting the lesser corner of the hyoid bone, thus separating the hyoid bone from the myelohyoid. This permits an additional several centimeters of mobilization to be achieved. All in all, a resection of one-half of the entire trachea, i.e., about 6 cm, can be achieved with primary reconstruction.

The anesthesia of patients with trachea obstruction is critical. Cardiopulmonary bypass has been used for tracheal surgery but is not needed for most cases. The endotracheal tube should be passed beyond the level of the lesion, and this may at times require dilation or coring out the tumor with a rigid bronchoscope. The trachea is transected distal to the lesion so that an airway can be introduced across the operative field directly into the distal trachea. Lesions in the upper half of the trachea that are known to be benign are best approached using a cervical collar incision. It may be necessary to prepare for a cervical mediastinal and possibly thoracic approach as well. If more exposure than can be achieved with a cervical approach is necessary, then the collar incision should be extended down to a partial median sternotomy. The upper sternum is split for a part of its length and horizontal division of the sternum is performed into the interspace. The incision sometimes needs to be extended through the fourth interspace on the right to allow for mobilization of the hilus of the right lung to further release the trachea. This permits wide exposure of the entire trachea. Lesions in the lower half of the trachea are usually approached through a fourth interspace, right posterolateral thoracotomy.

RECONSTRUCTION OF THE TRACHEA

Dissection is performed close to the trachea to avoid injurying the recurrent nerves, especially near the cricoid cartilage. Direct isolation of the nerves is avoided. The trachea is approximated after resection while the neck is flexed by the anesthesiologist and this demonstrates whether further mobilization using the previously discussed techniques may be necessary. Interrupted sutures are placed, usually 4-0 Vicryl, although in some friable tissues 4-0 Prolene may be necessary on pledgets.

For reconstruction of the lower trachea, anatomic mobilization is accomplished prior to severing the trachea since this allows improved exposure and handling of the trachea. It is important to note that laryngeal release is not helpful in performing resection of the carina; however the other maneuvers, i.e., hilar release and neck flexion are extremely helpful. Figure 24.2a-b, 24.3a-b and 24.4a-b show surgical techniques for procedures involving the upper and lower trachea.

In a patient with adenoid cystic carcinoma, assessment of the tumor spread is

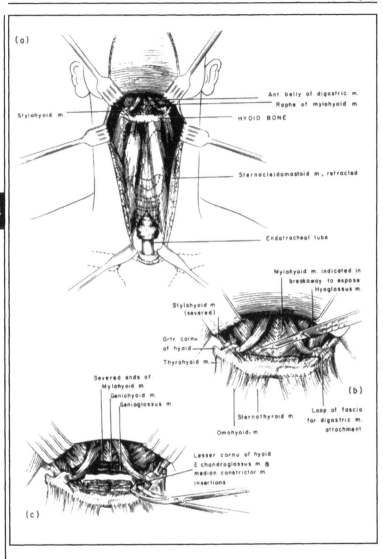

Fig. 24.1. Laryngeal release procedure to gain length for primary tracheal reconstruction. a) The anatomic relationships are shown. b) The muscles attached to the superior surface of the hyoid bone is transected. c) The lesser corner of the hyoid bone is palpated and transected, separating the hyoid bone from the myelohyoid. Reprinted with permission from: Suprathyroid Release for Tracheal Anastomosis. Arch Otolaryngol 99:256. ©1974 American Medical Association.

Fig. 24.2. Surgery of the upper trachea. a) The cervical incision may be extended to a partial median sternotomy of any length. b) Suture repair of the trachea after resection of the lesion. The prior endotracheal tube is pulled back and a flexible armored endotracheal tube is used to intubate the distal trachea across the operative field.

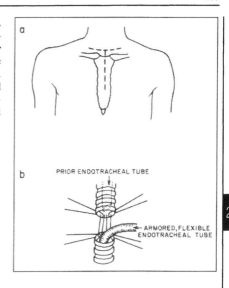

24

Fig. 24.3. Surgery of the lower trachea. a) Via a right posterolateral thoracotomy, exposure of the distal trachea is achieved. b) The trachea is transected and the mainstem bronchus intubated. The tracheal lesion is resected and primary anastomosis done.

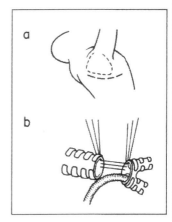

Fig. 24.4. Tracheal reconstruction after carinal resection. a) demonstrates the lines of resection.

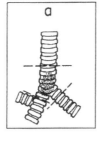

b) Demonstrates one type of reconstruction in which the right mainstem bronchus is sutured end-to-end to the trachea, and the left mainstem bronchus is sutured end-to-side to the trachea.

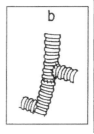

extremely important because, as mentioned, this tumor can spread for long distances submucosally. Hence, tissue is sampled 1-2 cm above and below the lesion as visible bronchoscopically. Contraindications to tracheal resection are cervical node metastases, esophageal or vena caval involvement, and Horner's syndrome. Relative contraindications may include laryngeal hemiparesis. These patients should undergo radiotherapy and laser obliteration of the tumor endoscopically. It should be noted that an additional technique, besides passing the endotracheal tube across the operative field into the distal trachea or mainstem bronchus, is to use flexible tubes through which high frequency jet ventilation is instituted. These small tubes can be passed down either mainstem bronchus and the operation can be continued in the trachea with minimal disturbance by the tubes.

TRACHEAL INNOMINATE FISTULA

This is an erosion injury of the innominate artery by a tracheostomy tube. The usual injury is from a low tracheostomy where the lesser curve of the endotracheal tube rests up against the innominate artery and causes direct erosion into the posterior portion of the artery. This results in exsanguinating hemorrhage up through the tracheostomy. It should be noted that much more rarely a tracheal innominate fistula is secondary to a correctly placed tracheostomy tube in which the balloon actually erodes into the innominate artery. These patients frequently have a herald bleed in which there is a small amount of blood coming from the tracheostomy rather than the fulminant exsanguinating hemorrhage of which these herald bleeds are a harbinger.

Management of a tracheal innominate fistula requires immediate attention. The immediate methods of achieving some control in the Intensive Care Unit prior to surgery is shown in Figure 24.5 a-d. The patient may be intubated from above, i.e., orotracheally and the tracheostomy tube removed. The cuff of the oral tracheal tube is then inflated up against the bleeding innominate artery. This achieves airway control and temporary control of the bleeding. Finger compression of the fistula is performed while the endotracheal tube is positioned and inflated over the fistula. Another method is to place a rigid bronchoscope and push the tracheostomy tube cuff up against the bleeding innominate artery. The first method is the preferred method for emergency control in the Intensive Care Unit. The patient should be immediately transported to the operating room where a median sternotomy is performed and the innominate artery ligated proximally and distally. No attempt is made to place a graft or otherwise revascularize the innominate artery because the erosion nearly always involves some amount of infected material from the trachea and reactive inflammation around the innominate artery. This lesion has a high mortality; however people have been saved by early aggressive management.

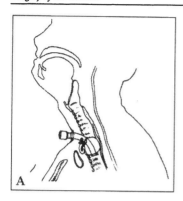

Fig. 24.5a. Immediate vascular control for tracheoinnominate fistula. In most cases, the inner curve of the tracheostomy has eroded into the innominate artery.

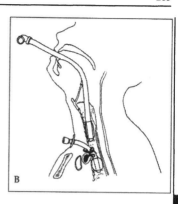

Fig. 24.5b. Intubation from above is performed.

24

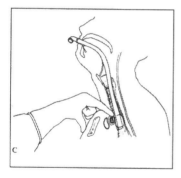

Fig. 24.5c. The tracheostomy tube is removed and finger compression of the fistula is done while the endotracheal tube is inflated over the fistula.

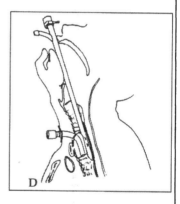

Fig. 24.5d. In another method, rigid bronchoscopy can be used to push the tracheostomy tube against the fistula to control hemorrhage.

TRACHEOESOPHAGEAL FISTULA

There are three varieties, acquired, malignant and congenital.

ACQUIRED NONMALIGNANT TRACHEOESOPHAGEAL FISTULA

Acquired nonmalignant tracheoesophageal fistulas may rise from pressure erosion at the site of an endotracheal tube. This is the most common cause at the present time and usually occurs in the presence of an esophageal feeding tube. However, in the past granulomatous mediastinal infections and trauma were more important. The incidence of this type of acquired tracheoesophageal fistula secondary to endotracheal cuff erosion is starting to decline because of the use of high volume low pressure cuffed endotracheal tubes. If the tracheoesophageal fistula occurs while the patient is still endotracheally intubated, reconstruction is not performed. Instead, the endotracheal tube is replaced with a low pressure cuff which is placed, if possible, as a tracheostomy below the level of the fistula. A gastrostomy may be performed to keep the stomach empty to prevent reflux, and a feeding jejunostomy is placed. The feeding tube through the esophagus is removed. When the patient is able to be extubated, the benign tracheoesophageal fistula is approached through a oblique sternocleidomastoid type incision on the left side. Another approach is a collar incision. A portion of the trachea usually needs to be resected with intubation of the distal trachea across the operative field or conversely using high frequency jet ventilation. The esophagus is repaired and then the trachea is reapproximated and re-anastomosed using 4-0 Vicryl in an interrupted fashion.

Intrathoracic tracheoesophageal fistulas are very rare, and the best approach is a right thoracotomy. The safest interposition is a intercostal muscle flap. For the cervical trachea, the sternal hyoid muscle is detached and sutured over the esophageal closure to an anastomosis of the trachea. Tracheal resection in the intrathoracic portion is usually not required since circumferential injuries with a cuff usually do not occur here. Sometimes additional tissue can be "stolen" away from the esophagus to close a large defect in the membraneous wall of the trachea, and this is perferred over a resection.

MALIGNANT TRACHEOESOPHAGEAL FISTULA

This usually results from malignant esophageal tumors eroding into the trachea. Eighty percent of these patients are dead from their cancer within 3 months, even when treated. It should be noted that all untreated patients, however, die miserably from repeated aspirations and some attempt at palliation is usually indicated. The two main treatment methods include esophageal intubation and surgical exclusion with bypass procedures. These accomplish the goals of excluding the fistula and restoring continuity of the esophagus. Esophageal intubation is effective for a period of time by sealing the fistula and relieving the obstruction temporarily. Diversion may be performed by cervical esophagostomy and gastrostomy. In esophageal exclusion with stomach or colonic bypass, the esophagus is stapled proximal and distal to the fistula and a bypass is performed using a substernal stomach or colon approach. Alternatively, the esophageal exclusion could be performed and a cervical esophagostomy performed.

The role of radiation therapy is limited to de novo tracheoesophageal fistulas. On the other hand, if the patient has already had radiation therapy, the chance of any success with further radiation is nil.

CONGENITAL TRACHEOESOPHAGEAL FISTULA

This is shown in Figure 24.6a-b. The most common variety involves the esophageal atresia proximally with a distal esophageal fistula. This occurs over 85% of

Fig. 24.6a. Incidence of the most common forms of tracheoesophageal fistula. A= 85%; B= 8%; C= 4%. Reprinted with permission from: Shields TW. General Thoracic Surgery. 4th ed. Fig. 79-5.

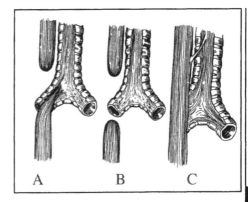

A B C

Fig. 24.6b. Standard initial management of the most common type of tracheoesophageal fistula with proximal esophageal decompression and gastrostomy in preparation for definitive surgery. Reprinted with permission from: Shields TW. General Thoracic Surgery. 4th ed. Fig. 79-2.

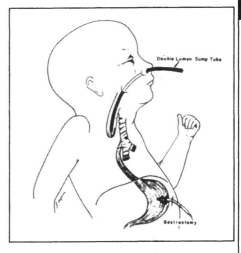

the time with other less common variants being the proximal and distal esophageal atresia without a fistula or the H-type fistula. When a congenital tracheoesophageal fistula is suspected, an orogastric tube should be passed and the diagnosis confirmed since the tube cannot be passed further than the end of the proximal pouch. Once the diagnosis is made of the most common variety of the tracheoesophageal fistula, the proximal pouch is decompressed with the sump catheter. The child is placed prone or with the head at 45° to prevent aspiration. It is important to establish the status of the child as an operative candidate according to the Waterston criteria. The Waterston A category is an infant with a birth weight greater than 2500 grams and no evidence of pneumonia and no major anomalies. Group B is an infant with a birth weight of 1800-2500 grams who may have pneumonia but has no major associated abnormalities. In group C, the infant is less than 1800 grams and does have major associated anomalies. Children who have Waterston A classification are candidates for immediate repair, at which time they undergo a right thoracotomy and primary repair of the trachea and esophagus with ligation and transection of the fistula. Children classified as Waterston B can be managed by delayed primary repair. There is decompression of the proximal esophagus with a sump catheter and the patient undergoes a gastrostomy to prevent further reflux into the trachea. Once the pneumonia has cleared, primary repair is performed. Children of the Waterston C are very difficult to manage and their treatment must be individualized, and the most serious life-threatening disease taken care of first if feasible. Sometimes it may be difficult to mobilize the proximal and distal esophagus enough to achieve primary repair. In this case an esophageal myotomy may be necessary to increase the length of the proximal pouch to relieve tension.

SUGGESTED READING

1.   Grillo HC. Surgery of the trachea. In: Current Problems in Surgery. Chicago: Year Book Medical Publishers, 1970.
2.   Perelman MI, Korolevo N. Surgery of the trachea. World J Surgery 1980; 4:583.
3.   Grillo HC. Tracheal tumors. Surgical management. Ann Thoracic Surg 1978; 26:112.
4.   Pearson FG, Todd TRJ, Cooper JD. Experience with primary neoplasms of the trachea and carina. J Thoracic Cardiovasc Surg 1984; 88:511-518.
5.   Grillo HC. Surgical management of post-intubation tracheal injuries. J Thoracic Cardiovasc Surg 1979; 78:860-875.

Thoracoscopy

Fritz J. Baumgartner

Video assisted thoracoscopic surgery (VATS) is a new technique in which standard procedures are performed utilizing a video camera in a much less invasive manner than a standard thoracotomy. The advantage of this approach is a marked decrease in postoperative pain because of elimination of the standard thoracotomy.

An important principle to keep in mind is that the actual procedure should be the same as if it were performed with an open technique, i.e., the operation must not be compromised to suit the limits of endoscopic surgery. There have been reports, for example, of limited thoracoscopic resection of lung cancer resulting in early recurrence. One must never compromise the care of the patient to suit the technique.

The procedures which are particularly suited to thoracoscopy include apical bleb resection and pleural abrasion, lung biopsy, cancer and node biopsy, sympathectomy and esophagomyotomy.

25

GENERAL PRINCIPLES

VATS requires both basic instruments and techniques which have been adapted from laparoscopic technique. Generally, the patient is placed in a lateral decubitus position, although the position may vary depending on the exact location of the lesion. General anesthesia is utilized and a double lumen endotracheal tube is required. Single lung ventilation is a prerequisite to permit adequate visualization of the thoracic cavity. Unlike laparoscopy where a pneumoperitoneum is induced, visualization of the chest is usually achieved by collapsing the ipsilateral lung rather than depressing it via an induced pneumothorax.

There are instances in which a single lumen endotracheal tube can be utilized. The pleural space is insufflated with CO_2 to depress the lung, making visualization possible. Initially the pressure not exceed 15 mmHg but can be raised to 20 mmHg with close hemodynamic observation. Tension pneumothorax can lead to electromechanical dissociation, particularly on insufflation of the left hemithorax.

A 10 mm Olympus thoracoscope is used with a camera adapted to a video monitor. The video monitor should be positioned on the opposite side of the surgeon. The surgeon generally stands posterior to the patient and the video monitor is anterior. It is preferred to have two video monitors, one on either side of the patient. The video monitor should be in a straight line drawn between the surgeon, the lesion and the monitor. Ports are generally beneficial for instrument placement, although this is not absolutely required. Generally, these thoraco-ports are most useful for inserting the camera since placing the camera directly through an incision may smear the lens and produce a poor image.

The other trocars used are: The 5 mm trocar for the instruments although frequently these can be left out and the instruments placed directly through the incision in the chest wall. A larger 15 mm trocar is used for placement of the stapling device.

A biopsy forceps, scissors or electrocautery/suction/irrigator are the primary tools used in thoracoscopy. For a general overview of the thoracic cavity, the best trocar placement for initial inspection is the midaxillary to posterior axillary line in the sixth intercostal space. This gives good visualization of the entire thoracic cavity. It is important after making the incision to dissect down into the pleural space and use blunt dissection to enter the pleural cavity after the lung has been deflated in a similar way as for insertion of a chest tube. This is to avoid injury to the lung should the lung be stuck up against the chest wall. Another helpful hint is that, with the patient placed in the lateral decubitus position, the table should be bent such that the lower extremities are brought down to open up the intercostal spaces. This will help in placing the trocars.

The technique of thoracoscopy requires a new set of manual skills which, although applicable from the open technique, still need to be refined in terms of hand/eye coordination. Several useful hints are mandatory in allowing for a smooth operation.

1. The trocar sites and thoracoscope should be placed far away from the lesion. Initially, it was thought that placing the camera directly over the lesion permitted the best view. This, however has since been found to be exactly the wrong approach. Placing the thoracoscope far away from the lesion allows for a panoramic view of the entire thoracic cavity and allows for evaluation of the lesion with respect to surrounding structures.

Figure 25.1 shows the relationship of the thoracoscope and various instruments to the lesion in terms of the "baseball diamond concept". In this model, the surgeon with the thoracoscope stands at home plate. The lesion is at second base, far away from home plate, and the various instruments including dissector, scissors, cautery/suction/irrigator are at first and third bases. Generally, the surgeon uses both hands to manipulate the instruments at first and third base while the assistant holds the telescope which is positioned far away from the lesion.

2. Instruments should be placed away from the lesion and pointed in the same general direction, i.e., toward the lesion and in the same direction as the telescope. This is to avoid the phenomenon called mirror imaging, in which the instrument is placed towards the camera resulting in an extremely awkward situation for performing surgery, as if watching one's actions in a mirror. Placing all instruments in the same direction with the thoracoscope avoids this major problem.

3. The instruments should not be crowded together, lest the phenomenon of "fencing" ensues with the instruments clanging against one another.

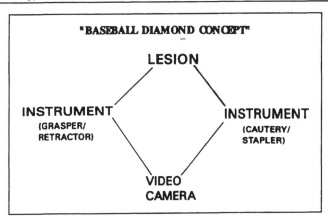

Fig. 25.1. "Baseball Diamond Concept" for video thoracoscopy. The video camera is positioned farthest from the lesion.

4. Random motion should be kept to a minimum. The surgeon should delegate which motions should generally be performed. The assistant should hold the camera and follow the movement of the surgeon without being told this continually. The team should manipulate only one instrument at a time to prevent chaos between the different instrument manipulators. Using a stepwise, systematic approach leads to a much more satisfactory operation.

APICAL BLEB RESECTION

Apical bleb resection is the simplest procedure and is as suited to thoracoscopic intervention as cholelithiasis is suited to laparoscopic intervention. Apical blebs are identified thoracoscopically by placing the camera in the sixth intercostal space in the midaxillary line. The grasper and the stapler are positioned at the first base and third base positions as discussed previously. Figure 25.2 describes the positions.

Sometimes the apical bleb is very difficult to identify and a 30° scope may be useful instead of the 0° scope in going up above the apex in the lung to identify the lesion. It may be helpful to instill some saline in cases in which no bleb is identified and try to inflate the lung to see if there is any air coming out of the apical bleb. After the bleb has been resected, a pleural abrasion procedure should be done. This is accomplished by taking a piece of cautery scratch pad and inserting it through the stapler trocar site using a grasper and rubbing the apical parietal surface of the pleural space to produce ultimate fusion of the visceral and parietal areas.

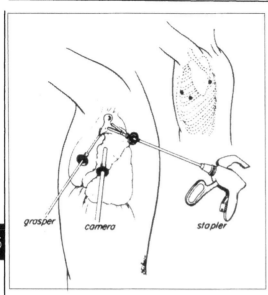

Fig. 25.2. Video thoracoscopy for apical bleb resection and abrasion pleurodesis. The camera is inserted into the sixth interspace in the mid-axillary line. The grasper and stapler are on either side. Abrasion pleurodesis is performed using a cautery scratch pad inserted through the stapler part. Reprinted with permission from: Ann Thoracic Surg 1992; 54:800-807.

grasper   camera   stapler

LUNG BIOPSY

This is another fairly easy surgical maneuver to perform using video thoracoscopy. The trocar position depends on the location of the lesion. The lesion of the superior segment of the lower lobe is depicted in Figure 25.3. The grasper and staple gun can be positioned on opposite sides of the camera.

The thoracoscopic pulmonary mass resection is generally only applicable for lesions located in the periphery of the lung that can be visualized. It is very difficult to achieve the tactile sense thoracoscopically that is second nature for the surgeon in an open chest case. Nonetheless, there are techniques whereby lesions not visible on the visceral surface of the lung can be resected. One method which has been utilized at our hospital is to perform preoperative CT-guided needle localization of such masses, injecting methylene blue at the same time. Using the needle localization/methylene blue technique, one is then able to identify the lesion quite readily using thoracoscopy and can staple resect the lesions. It is important to note that this procedure should never be used to perform a resection of lung cancer which may compromise the surgical outcome. In other words, if cancer is identified on wedge resection, then a formal pulmonary lobectomy should be performed. Formal lobectomy has been shown to decrease the recurrence of primary lung carcinoma, as compared to wedge resections or segmentectomies. Until prospective randomized studies document no survival disadvantage with wedge resection, the operation cannot be recommended.

Fig. 25.3. Video thoracoscopy for resection of a lesion in the superior segment of the lower lobe. Reprinted with permission from: Ann Thoracic Surg 1992; 54:800-807.

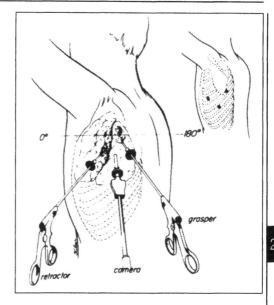

25

CANCER STAGING AND LYMPH NODE BIOPSY

Using video thoracoscopy, cancer staging can be accomplished by visual inspection of the pleural space. If the pleura is studded with tumor, a diagnosis of stage IV lung carcinoma can be made and the patient can be considered unresectable. Lymph node biopsies can be performed in locations such as the aortopulmonary window or subcarinal regions, as well as hilar and tracheobronchial areas. This approach, in addition to the grasper and scissors on either side of the camera, utilizes a retractor to pull down the apex of the lung to permit improved visualization of the area. Although this is not necessary, it is helpful. This can be accomplished using a fan retractor or other device.

DORSAL SYMPATHECTOMY

Dorsal (thoracic) sympathectomy for hyperhidrosis or causalgia (posttraumatic vasomotor dystrophy) can be accomplished successfully using video-thoracoscopic techniques (see Chapter 22). The sympathetic chain is located paravertebrally in the posterior mediastinum and is easily accessible by endoscopy. Although double-lumen endotracheal intubation and the lateral decubitus position are reliable with excellent results, we prefer the supine position with single-lumen endotracheal intubation and insufflation of the pleural space with CO_2. Intrathoracic pressures of up to 15 mmHg are generally well-tolerated for short periods of time, but higher pressures, especially in the left hemithorax, should be carefully monitored hemodynamically for tension pneumothorax-related electromechanical dissociation (EMD). The endoscope is inserted into the pleural space through the 4th interspace in the anterior axillary line. Another port is placed at the base of the axillary hairline and a spatula cautery is used to cautery-transect the sympathectic chain at the level of the 2nd and 3rd ganglia. Reverse Trendelenburg and rotation of the patient away from the side of surgery facilitates exposure. Generally the 1st ganglion is avoided to prevent any danger of Horner's syndrome.

ESOPHAGOMYOTOMY

This can be successfully performed thoracoscopically although it is the most difficult and technically challenging procedure thus described. A truncal vagotomy may also be performed thoracoscopically using similar techniques. The critical concept is that the esophagus must be accurately and easily identified. This is best accomplished by insertion of an upper GI flexible endoscope. We have used an esophagoscope with a flashing light which can be transmitted through the esophagus to permit even easier identification. A second major function of esophagoscopy is to identify at what point the myotomy is complete. Once the esophagomyotomy is carried down onto the surface of the stomach (where classically the longitudinal venous plexus becomes transverse), then transection of the lower esophageal sphincter has occurred. At this point on endoscopy, one will see a sudden relaxation which yields evidence that the esophagomyotomy is complete.

Besides the thoracoscope port, the other access ports include a lung retractor, scissors attached to an electrocautery for the esophagomyotomy, a separate retractor for the diaphragm and a grasper to pick up the esophagus.

The thoracoscope is inserted in the fourth intercostal space of the posterior axillary line and directed down toward the diaphragm and lower esophagus. It is important to note that the video screen should be positioned at the feet of the patient, since the surgeon will be standing near the head of the patient. The lung is deflated and the patient placed with the head down to permit better retraction of the lower lobe cephalad. A lung retractor is placed in the fourth intercostal space in the anterior axillary line to retract the lower lobe anteriorly and superiorly. Another retractor is placed in the seventh intercostal space in the posterior axillary line to push the diaphragm inferiorly. A grasper and scissors are placed in the sixth to eighth intercostal space more posteriorly to manipulate and myotomize the esophagus. The thoracoscopic camera should be placed through the seventh intercostal space quite posteriorly to permit best visualization of the esophagus. The port positions are shown in Figure 25.4.

Outstanding results have been achieved using thoracoscopic technique for esophagomyotomy. We have successfully performed the procedure at Harbor. The pain and disability of a thoracotomy is one reason why the less invasive but less effective pneumatic dilatation is so frequently used as the initial management. Pneumatic dilatation even in the best of hands has only an 80% initial success rate and has its own complications, including esophageal perforation. The advantage

Fig. 25.4. Video thoraco-
scopic esophagomyotomy.
The thoracoscope is placed
into the seventh interspace
at the midaxillary line. The
grasper, scissors and retrac-
tors are placed via other
ports. Esophagoscopy from
above helps visualize the
esophagus (see text). Re-
printed with permission
from: Am Thoracic Surg
1992; 54:800-807.

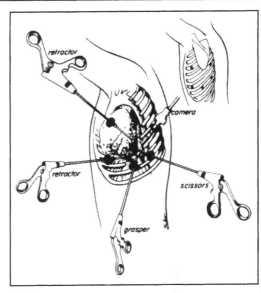

25

of this video guided approach is that it avoids the more invasive thoracotomy.

It should also be noted that esophagomyotomy alone remains controversial as there are some groups that routinely perform fundoplication with their esophagomyotomies to prevent gastroesophageal reflux. Most groups, however, feel that fundoplication is not necessary as gastroesophageal reflux should not be a problem if the lower extent of the esophagomyotomy is limited to the lower esophageal sphincter. The myotomy should only extend onto the surface of the stomach to the extent determined by the endoscopist seeing a sudden relaxation of the lower esophageal sphincter. Laparoscopic esophagomyotomy without fundoplication is conceptually sound.

SUGGESTED READING

1. Landreneau RJ, Mack MJ, Hazelrigg SR et al. Video-assisted thoracic surgery: Basic technical concepts and intercostal approach strategies. Ann Thorac Surg 1992; 54:800-807.

2. Miller DL, Allen MS, Deschamps C et al. Video-assisted thoracic surgical rocedure: Management of a solitary pulmonary nodule. Mayo Clin Proc 1992; 67:462-464.

3. Pellegrini C, Wetter LA, Patti M et al. Thoracoscopic esophagomyotomy. Ann Surg 1992; 216:291-299.

A

Printed and bound by CPI Group (UK) Ltd, Croydon, CR0 4YY

23/10/2024

01777672-0001